ML

STRATHCLYDE UNIVERSITY LIBRARY

30125 00366418 1

KV-577-615

ᴈ returned on ᴄ
ᴊate stamped below.

3Hῆ: Oᴛ

ANDERSONIAN LIBRARY
✳
WITHDRAWN
FROM
LIBRARY
STOCK
✳
UNIVERSITY OF STRATHCLYDE

MICROSTRUCTURAL SCIENCE FOR THIN FILM METALLIZATIONS IN ELECTRONIC APPLICATIONS

1994527

MICROSTRUCTURAL SCIENCE FOR THIN FILM METALLIZATIONS IN ELECTRONIC APPLICATIONS

Proceedings of the Topical Symposium held at the Annual Meeting of The Minerals, Metals & Materials Society, sponsored by the Electronic Materials Device Committee of TMS, at Phoenix, Arizona, January 26-27, 1988.

Edited by

John Sanchez
Lawrence Berkeley Laboratory
University of California, Berkeley, California

David A. Smith
IBM
Yorktown Heights, New York

Nimal DeLanerolle
Standard Microsystems Corporation
Hauppage, New York

A Publication of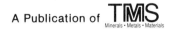

A Publication of The Minerals, Metals & Materials Society
420 Commonwealth Drive
Warrendale, Pennsylvania 15086
(412) 776-9024

The Minerals, Metals & Materials Society is not responsible for statements or opinions and absolved of liability due to misuse of information
contained in this publication.

Printed in the United States of America
Library of Congress Catalog Number 88-62442
ISBN Number 0-87339-077-6

Authorization to photocopy items for internal or personal use, or the internal or
personal use of specific clients, is
granted by The Minerals, Metals &
Materials Society for users registered
with the Copyright Clearance Center
(CCC) Transactional Reporting Service,
provided that the base fee of $3.00 per
copy is paid directly to Copyright Clearance Center, 27 Congress Street, Salem,
Massachusetts 01970. For those organi-
zations that have been granted a photocopy license by Copyright Clearance
Center, a separate system of payment
has been arranged.

Minerals • Metals • Materials

© 1988

D
669.95
MIC

PREFACE

This volume contains the abstracts, extended abstracts, and critically reviewed manuscripts representing the papers presented at the Topical Symposium "Microstructural Science for Thin Film Metallizations in Electronics Applications", held at the 1988 Annual Meeting of The Minerals, Metals & Materials Society at Phoenix, Arizona, January 26-27, 1988. This Topical Symposium was sponsored by the Electronic Materials Device Committee (EDMC) of The Minerals, Metals & Materials Society.

The overall focus of the two-day Topical Symposium was to emphasize the importance of microstructural descriptions and characterizations for the proper understanding of properties of thin solid film systems. The determination of fundamental microstructure->properties relationships is the classical metallurgical/materials science approach to materials characterization and design. Thus, the papers presented at the Topical Symposium were chosen to concentrate on descriptions of general microstructural features of thin films, and on those microstructure-> properties relationships relevant to the general electronics community.

Topics presented in Session I, "Requirements and Deposition Methods", demonstrated the breadth of electronics applications for thin films, such as high T_c thin film superconductors, metal oxide diffusion barriers in integrated circuits, chemical vapor deposition (CVD) modelling, CVD of metals, and metal-GaAs reactions. Session II, "Microstructure of Thin Films", concentrated on metallurgical descriptions of microstructure. It was shown how anisotropic surface and interfacial energies may determine deposited film crystallographic texture. The theory, modelling, and experimental evidence for the evolution of microstructure as a function of deposition and processing conditions were presented since grain size, grain growth, and the grain size distribution often play the most critical role in determining film properties. Topics in Session III, "Microstructure->Property Relationships and Failure Mechanisms", focussed on the origin of mechanical stress and its effects in thin films, and on electromigration mechanisms in thin film interconnects. Origins of stresses in thin films that were presented include mechanisms operative during film deposition or growth, thermal expansion mismatch between film and substrate, grain growth, void growth or shrinkage, and electromigration induced mass-flux divergence. During post-deposition thermal processing, stresses may be relieved via plastic deformation or significantly altered (increased or decreased) via microstructural changes within the thin film. Film-substrate delamination and cracking are possible under certain microstructure and processing conditions. Despite active effort for over 20 years, electromigration induced open and short circuit failures continue to compromise the lifetime reliability of integrated circuits. Grain size, grain size uniformity, solute distribution, and local temperature gradients critically determine the electromigration mass flux leading to circuit failure. The consensus was that, while much progress has been made, a large amount of work is necessary in order to provide a comprehensive microstructurally based model for electromigration damage and failure.

Thin film materials systems are central to electronics, microelectronics, magneto-optical, and electro-optical technologies. The 19 papers presented here and the well-attended sessions indicate the growing recognition for and the increasing application of microstructural characterizations and microstructure-> properties relationships as new materials are designed for these emerging and maturing technologies.

The assistance of the staff of The Minerals, Metals & Materials Society, the sponsorship of EDMC, and the enthusiastic participation of the invited speakers, authors and session attendees is gratefully acknowledged. The efforts of the manuscript reviewers which greatly assisted in the editorial preparation of this Proceedings are also acknowledged.

The Organizing Committee

John Sanchez, Co-Chairman
David A. Smith, Co-Chairman
Nimal DeLanerolle

TABLE OF CONTENTS

Session I:

REQUIREMENTS AND DEPOSITION
METHODS FOR THIN FILMS

Session Chairman
Nimal DeLanerolle
Standard Microsystems Corp.
Hauppage, New York

SYNTHESIS AND PROPERTIES OF HIGH T_C THIN FILM SUPERCONDUCTORS

J. C. Bravman and R. W. Barton,
for the Stanford High T_C Thin Films Group

Department of Materials Science and Engineering
Stanford University
Stanford, California

Abstract

In this Extended Abstract we briefly describe our recent results concerning the synthesis and properties of thin films of the new class of ceramic oxide superconductors. To date most of our work has centered on the Y–Ba–Cu–O system, for which $T_C \approx 90°K$. Films with sharp resistive transitions ($\Delta K < 2°K$) and high critical currents ($J_C > 10^6$ amps/cm^2 at 4.2°K) have been fabricated by sputtering, electron beam evaporation and molecular beam epitaxy.

Microstructural Science for Thin Film
Metallizations in Electronics Applications
Edited by J. Sanchez, D.A. Smith and N. DeLanerolle
The Minerals, Metals & Materials Society, 1988

The discovery of superconducting materials with transition temperatures above the boiling point of liquid nitrogen has generated unprece-
dented levels of activity within the scientific and engineering communi-
ties. At Stanford (1) most our work involves the synthesis and characterization of *thin films* of the new superconductors, particularly those
based on the Y-Ba-Cu-O system. These materials were first shown by Chu
and co-workers to have transition temperatures above 90°K (2). Using
several methods we are able to grow both polycrystalline and near-single
crystal thin films over a wide range of compositions. These films have
already demonstrated critical current densities at low fields as high as
10^7 amps per square centimeter (measured at 4.2°K), well in excess of
those achieved in even the best sintered materials. Full characterization of our materials includes a variety of transport measurements
(resistivity, Hall effect, magnetoresistance, J_c, etc.) and physical
examinations (x-ray diffraction, transmission electron microscopy, photoelectron spectroscopy, etc.).

We have explored three different methods for growing thin films of
these materials: magnetron sputtering (3), electron beam evaporation
(4,5), and, in collaboration with a group at Varian Associates, molecular beam epitaxy (6). Each has proven capable of producing films with
sharp transition temperatures at or above 90°K and with very high critical currents. The techniques share much in common. First, each makes use
of elemental metal sources (in the MBE case, dysprosium is substituted
for yttrium, given the difficulty in using the latter in an MBE system).
Evaporation from oxide sources (CuO, BaO and Y_2O_3) was not successful
(4). Second, a high temperature anneal in an oxidizing ambient is
required to transform the as-deposited films, which are amorphous or
highly disordered, into superconductors; oxygen stoichiometry is generally thought to be an important parameter in these films (7). And third,
single crystal {100} $SrTiO_3$ is used as the substrate. Earlier attempts
to use single crystals of Al_2O_3, Si, MgO and ZrO_2, as well as amorphous
SiO_2, resulted in the growth of poorer quality films (lower and/or
broader transition temperatures and lower critical currents). Using
$SrTiO_3$ substrates we can deposit highly oriented films which have either
the a-axis or the c-axis perpendicular to the substrate (overlap with
substrate peaks makes it difficult to distinguish between a-axis and b-
axis oriented films via x-ray techniques, and thus what we call an a-
axis film may contain both orientations (5)). Control over which orientation predominates can, to a certain degree, be established through
changes in the film stoichiometry. These changes can lead, however, to
the incorporation of second phase precipitates which in general are
deleterious to the superconducting properties of the films. It is
thought that segregation of non-superconducting phases to the grain
boundaries is particularly harmful, although the relative importance of
inter-grain vs. intra-grain properties has not been fully established.
In the period ahead one major task will be to establish full control
over the microstructure of the films as a function of processing parameters. Another area of great interest is that of *in-situ* processing (8),
in which the oxygen is incorporated in the film *during* deposition. The
approach here would be to include an oxygen ion gun into one or more of
our deposition chambers; our hope is to produce a more active oxygen
species.

4

A recent discovery made at Stanford is that of a new ordered defect structure, $Y_2Ba_4Cu_8O_{20-x}$ (9). This differs from the standard $YBa_2Cu_3O_{7-x}$ phase by the inclusion of extra Cu-O planes between the two Ba-O planes. This material shows a resistive transition near 80°K and can be made nearly 95% pure. The properties of this phase are now being studied.

References

1. Members of the Stanford High T_C Thin Film Group are listed as authors in Refs. 3, 4, 5, 6 and 8; the first three authors in Ref. 6 are with Varian Inc.

2. C.W. Chu, P.H. Hor, R.L. Meng, L. Gao, Z.J. Huang and Y.Q. Wang, Phys. Rev. Lett., 58 (1987) 405.

3. K. Char, A.D. Kent, A. Kapitulnik, M.R. Beasley and T.H. Geballe, "Reactive magnetron sputtering of thin film superconductor $YBa_2Cu_3O_{7-x}$," App. Phys. Lett.,51(17) (1987) 1370-2.

4. M. Naito, R.H. Hammond, B. Oh, M.R. Hahn, J.W.P. Hsu, P. Rosenthal, A.F. Marshall, M.R. Beasley, T.H. Geballe and A. Kapitulnik, "Thin film synthesis of the high T_C oxide superconductor $YBa_2Cu_3O_{7-x}$," J. Mat. Res., 2(6) (1987)713-25.

5. B. Oh, M. Naito, S. Arnason, P. Rosenthal, R. Barton, M.R. Beasley, T.H. Geballe, R.H. Hammond and A. Kapitulnik, "Critical current densities and transport in superconducting $YBa_2Cu_3O_{7-x}$ films made by electron beam co-evaporation," App. Phys. Lett., 51(11) (1987) 852-4.

6. C. Webb, S.L. Weng, J.N. Eckstein, N. Missert, K. Char, D.G. Schlom, E.S. Hellman, M.R. Beasley, A. Kapitulnik and J. S. Harris, "Growth of high T_C superconducting thin films using molecular beam epitaxy techniques," App. Phys. Lett., 51(22) (1987) 1191-3.

7. R. Beyers, G. Lim, E.M. Engler, V.Y. lee, M.L. Ramirez, R.J. Savoy, R.D. Jacowitz, T.M. Shaw, S. La Place, R. Boehme, C.C. Tsuei, S.I. Park, M.W. Shafer and W.D. Gallagher, "Annealing treatment effects on structure and superconductivity in $YBa_2Cu_3O_{7-x}$," Appl. Phys. Lett., 51(8) (1987) 614-16.

8. A.F. Marshall, R.W. Barton, K. Char, A. Kapitulnik, B. Oh, R. Hammond and S.S. Laderman, "An ordered defect structure in epitaxial $YBa_2Cu_3O_{7-x}$ thin films," Phys. Rev. B, in press.

9. D. K. Lathrop, S. E. Russek and R. A. Buhrman, "Production of $YBa_2Cu_3O_{7-x}$ superconducting thin films in-situ by high-pressure reactive evaporation and rapid thermal annealing," Appl. Phys. Lett., 51(19) (1987) 1554-6.

CHEMICAL VAPOR DEPOSITION OF METALS*

Robert S. Blewer

Center For Radiation Hardened Microelectronics
Sandia National Laboratories
Albuquerque, New Mexico 87185

Abstract

Dramatic advances in selective and non-selective low pressure CVD techniques have been reported in recent months. Particularly in the selective deposition of refractory metals and metal silicides, several important new developments have stimulated strong interest within the microelectronics community. Fundamental studies have revealed the importance of byproducts on deposition rate and degree of selectivity, while new reactor approaches have led to results that appear to meet manufacturability requirements for advanced VLSI designs. New process gas mixtures have been used to achieve deposition rates which exceed 1μm/min. for selective tungsten films. Reliable refractory metal gate technology and multi-active-level devices have been demonstrated using CVD tungsten and/or molybdenum and tungsten silicide. A survey of these results and a perspective on the current status of CVD of metals for VLSI applications will be presented.

*This work was performed at Sandia National Laboratories by the US Department of Energy under contract DE-AC04-76DP00789.

Microstructural Science for Thin Film
Metallizations in Electronics Applications
Edited by J. Sanchez, D.A. Smith and N. DeLanerolle
The Minerals, Metals & Materials Society, 1988

MODELING OF CHEMICAL VAPOR DEPOSITION REACTORS

A. Sherman

Varian Associates, Inc.
Palo Alto, CA   94303

Abstract

The design and development of chemical vapor deposition reactors has historically been done experimentally. In recent years, however, the performance of these reactors has been pushed to the limit, requiring more time and expense to develop new systems. This paper will review the recent development of mathematical techniques for the modeling of such reactors. Methods for modeling of low-pressure, cold-wall and hot-wall systems will be covered as well as high-pressure reactors ($\approx$ atmospheric). The basic concepts involved in describing the flow of reacting gases, including gas phase and surface reactions, will be presented. The techniques that have been employed (which include surface reactions in the mathematical description so that deposition rates may be calculated) will be reviewed as well.

Microstructural Science for Thin Film
Metallizations in Electronics Applications
Edited by J. Sanchez, D.A. Smith and N. DeLanerolle
The Minerals, Metals & Materials Society, 1988

Introduction

In recent years, integrated circuit designers have reduced circuit features to such small dimensions (i.e., 1 μm or less), that the performance of the chemical vapor deposition (CVD) reactors that are used to deposit thin films on these circuits is now being pushed to their limits. As one example, where older circuits could tolerate a variation of film thickness of ± 5% over a 100-mm diameter wafer, newer circuits are looking for ± 1-2% over 150-mm wafers. There has even been interest in using 200-mm wafers. As a result, the familiar experimental approach to the design of such reactors has become increasingly more difficult and expensive.

One way out of this difficulty is to set up a mathematical model of the CVD reactor, which can then be solved on available high-speed computers. Provided that the model is an accurate representation of the reactor, one can then design systems on the computer. In this way, we would be able to vary geometry, inlet and outlet flow configurations, as well as reactant gas mixtures and flow conditions on paper to determine the optimum reactor design.

The study of the fluid dynamic behavior of reacting gas flows is not a new subject. In the 1950's, such gas flows about a reentry nose cone for ballistic missile weapons was the subject of intense study. At the same time, the flow of combustion gases through rocket exhaust nozzles was being analyzed. Following this, interest in magnetohydrodynamic power generation [1] led to further study of such flows as well as their behavior when ionized and in the presence of electric and magnetic fields.

With the recent application of CVD techniques to thin film depositions for the integrated circuit industry, interest in reacting gas flows has continued. Initially, attention was focused on the atmospheric-pressure, cold-wall CVD reactor used to deposit epitaxial silicon thin films [2-4]. Later, studies were done of the hot-tube, low-pressure CVD reactor used primarily for polysilicon, SiO_2 and Si_3N_4 films [5,6]. More recently, detailed flow calculations have been reported for single-wafer, cold-wall reactors that can be used to deposit compound semiconductors from metal organic reactants (MOCVD) at high pressures (75 Torr - 1 atm.) or conducting refractory metal films at low pressure ($\approx$ 300 mTorr) [6-9].

Obviously, these studies included surface reactions in order to predict deposition rates. However, since the gas phase kinetics have been poorly understood, most used highly simplified kinetic schemes. Recent studies by the Sandia group has, however, examined a very complex system of possible reactions [10,11], and predicted species concentrations in the gas phase that have been confirmed by experimental measurements [12,13].

It will be the objective of the present paper to review this relatively new and complex field and to identify the applicability of the various techniques developed. Attention will be limited, however, to those CVD reactors where fluid flow plays an essential role, so low-pressure, hot-wall systems will not be considered in this review.

Governing Equations

Provided that the mean-free path in the reacting gas is much smaller than the characteristic length of a reactor of interest, a continuum description of the flow in a CVD reactor is appropriate. The governing equations are the Navier-Stokes equations for reacting gas mixtures, which allow for transport effects due to viscosity, thermal conductivity, and diffusion [11].

Overall mass continuity:

$$\frac{\partial \rho}{\partial t} + \frac{\partial (\rho u_i)}{\partial x_i} = 0 \tag{1}$$

where
ρ = mass density
t = time
u_i = velocity components
x_i = coordinates.

Conservation of momentum:

$$\frac{\partial}{\partial t}(\rho u_i) + \frac{\partial}{\partial x_i}(\rho u_i u_j) = -\frac{\partial p}{\partial x_i} + \frac{\partial \tau_{ik}}{\partial x_k} + \rho g_i \tag{2}$$

where
p = pressure
g_i = gravitational acceleration
τ_{ij} = stress tensor
and

$$\tau_{ij} = -\frac{2}{3}\mu \delta_{ij}\frac{\partial u_e}{\partial x_e} + \mu\left(\frac{\partial u_i}{\partial x_j} + \frac{\partial u_j}{\partial x_i}\right) \tag{3}$$

where μ = fluid viscosity.

11

It should be noted that this form of the stress tensor is only strictly valid for monatomic gases [14].

Conservation of energy:

$$\rho c_p \frac{\partial T}{\partial t} + \rho c_p u_i \frac{\partial T}{\partial x_i} = \frac{\partial}{\partial x_i}\left(\lambda \frac{\partial T}{\partial x_i}\right) + \frac{\partial p}{\partial t} + u_i \frac{\partial p}{\partial x_i}$$

(4)

$$+ \tau_{ij}\frac{\partial u_i}{\partial x_j} - \sum_{k=1}^{K} \rho Y_k V_{k_i} C_{p_k} \frac{\partial T}{\partial x_i} - \sum_{k=1}^{K} \omega_k h_k w_k$$

(5)

where

	c_p	=	specific heat at constant pressure
	T	=	temperature
	λ	=	thermal conductivity
	Y_k	=	mass fraction of k'th species
	V_{k_i}	=	diffusion velocity of k'th species
	ω_k	=	production or destruction rate of species k
	h_k	=	enthalpy of species k
	w_k	=	molecular weight of species k
	K	=	total number of species in gas

Species mass conservation:

$$\frac{\partial \rho_k}{\partial t} + \frac{\partial}{\partial x_i}(\rho_k u_i) = - \frac{\partial}{\partial x_i}(\rho Y_k V_{k_i}) + \omega_k w_k \qquad k = 1, K$$

(6)

where ρ_k = mass density of species k

$$V_{k_i} = \upsilon_{k_i} - u_i$$

(7)

υ_{k_i} = absolute velocity of species k

and where the diffusion velocity can be expressed as

$$V_{k_i} = -\left[\frac{D_k}{X_k}\frac{\partial X_k}{\partial x_i} - \frac{D_k^T}{X_k}\frac{1}{T}\frac{\partial T}{\partial x_i}\right]$$

(8)

where D_k =   diffusion coefficient of species k in the mixture

              X_k =   mole fraction of species k

              D_k^T =   thermal diffusion coefficient of species k.

Finally, we assume a perfect gas so that

$$p = \rho\frac{RT}{w}$$

(9)

where $\dfrac{R}{w}$ =   universal gas constant

                 =   average molecular weight of mixtures.

Reviewing these equations, we can make the following observations:

1.   Dependent variables p, ρ, T, Y_k, u_i

2.   Independent variables x_i, t

3.   Constants R, w_k

4.   Transport coefficients $\lambda, \mu, D_k, D_k^T = f(p,T,X_i)$

5.   Production/destruction rate $\omega_k = \omega_k(p,T,X_i)$

6. $c_p = c_p(T)$

Boundary Conditions:

      Flow - $u_i = 0$ on all solid surfaces
                   u_i specified at inlet and outlet to reactor.

Note that u_i is not strictly $= 0$ at a surface, which is either being ablated or deposited on. For typical CVD processes, this velocity is small enough to be neglected.

       Temperature -   on all surfaces bounding the flow, we can specify

                   T or $\partial T/\partial n$ and either can be a function of location on the boundary.

       Pressure -   p must be specified at all open boundaries (i.e., entrance and exit to reactor).

Mass Fractions -   At entrance open boundary, specify Y_k (unreacted gas mixture).

At exit open boundary, specify $\partial Y_k/\partial x_i = 0$ (no more change).

On solid surfaces, equate flux of species k to surface as described by continuum theory to flux as described by molecular theory.

$$\rho_k V_k = -\left(\sqrt{\frac{k_B T}{2\pi m_k}} \cdot \gamma_k\right) \rho_k v_{k,s}$$

(10)

where

γ_k = probability of reaction of species k on surface

$v_{k,s}$ = stoichiometric coefficient of solid bearing molecule k.

The end result is a mixed boundary condition involving Y_k and $\partial Y_k/\partial n$. If the molecule does not react with the surface, $\gamma_k = 0$ and the boundary condition reduces to $V_k = 0$. If thermal diffusion is neglected, then the condition reduces to $\partial Y_k/\partial n = 0$.

When the molecule reacts on the surface with unit probability, the assumption is often made that $Y_k = 0$. The correct boundary condition is given by Eqn. (6). The validity of assuming $Y_k = 0$ can be judged by looking at Eqn. (6). Assume $D^T_k = 0$ and consider an atmospheric pressure gas. Then,

$$\frac{1}{Y_k}\frac{dY_k}{dn} = \frac{\sqrt{\frac{k_B T}{2\pi m_k}}}{D_k} \cong 10^5$$

(11)

and

$$Y_{solid} = e^{\left[\ln Y_{gas} - 10^5 \cdot \delta\right]}$$

(12)

where we assume a concentration boundary layer of thickness δ and Y_{gas} is the mass fraction at its outer edge. Even for $\delta \approx 1$ mm, we still get $Y_{solid} \approx 0$. Note, however, that for low pressure CVD with $p \approx 100$ mTorr, that D_k will be very large and $Y_k = 0$ may not be an accurate boundary condition.

Finally, we should note that the species net production rate terms, $\dot{\omega}_k$, can be determined by defining the relevant gas phase reactions and using an Arrhenius form for the reaction rate constants.

Complexities and Simplifications

First, we must recognize that the system of equations governing the CVD reactor is highly nonlinear and extremely complex. Therefore, we cannot expect to solve them for the general case in three dimensions and unsteady in time. Aside from the considerable mathematical difficulty in solving these equations, even with the largest high speed computers, there is another more fundamental problem. The kinetic processes occurring in the gas phase and on the surface are not at all well understood. Probably the SiH_4 decomposition in the gas phase and Si deposition on hot wafer surfaces is best understood today, and the most complete analysis [11] includes 27 reactions and 17 species.

Therefore, it will be necessary to make many simplifying assumptions in order to achieve some success in modeling a CVD reactor. First, we note that all treatments to date have treated steady flows $(\partial/\partial t = 0)$. The next area of potential simplification is geometrical. A number of authors have looked at one-dimensional problems, two-dimensional problems within the boundary layer flow approximation, full two-dimensional problems, and even three-dimensional cases.

Aside from geometrical simplifications, a major reduction in complexity can be achieved by assuming the reactant is heavily diluted by a nonreactive gas. In this case, the transport properties (μ, λ) can be calculated as functions of p and T, but do not depend on Y_k's. At the same time, we can neglect the last two terms in the energy equation, Eqn. (3), that represent the diffusion of enthalpy and the heat created or lost by net production of species. In addition, since we are dealing with low Reynold's number flows in CVD reactors, we can also neglect heat generated by viscous dissipation, so we can drop the term $\tau_{ij} \partial u_i/\partial x_j$ from Eqn. (3).

With these assumptions, the momentum and energy equations are uncoupled from the species conservation equations, so that we can solve for the flow and temperature fields independently of the species fields. This allows a significant simplification of the numerical difficulties.

15

One-dimensional Flows

There are three flow geometries which can be studied in the one-dimensional flow limit. They are (1) impinging jet flow, (2) rotating disc flow, and (3) fully developed parallel plate channel flow. All have been studied for many physical problems, and they have recently been examined as limited models for CVD reactors.

For impinging jet and rotating disc flows, we visualize a disc of infinite radius and consider a reacting gas flow normal to it. For both these cases, a similar solution is possible if bouyancy forces are neglected. That is, a change of variables will reduce the original two-dimensional equations to a one-dimensional problem. The original variable transformation was suggested by von Karman [15]. A sketch of the general flow, with rotation, is shown in Fig. 1.

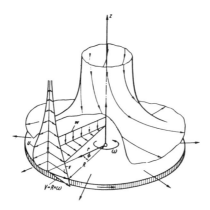

Fig. 1. Flow near a rotating disc [15].

To reduce the equations to the one-dimensional case, just assume that the velocities have the form:

$$u \propto rf(z), \quad v \propto rg(z), \quad w \propto h(z)$$

and, of course, for the impinging jet case $v = 0$.

Pollard and Newman [16] studied the one-dimensional rotating disc problem with gas phase and surface reactions, included enthalpy diffusion, and energy due to net species production in the energy equation but neglected thermal diffusion. They considered the case of deposition of silicon from $SiCl_4$ in H_2. The same equations were solved by Kee et al.[17] for silicon deposition from SiH_4 in H_2 or He. Their kinetic model included 27 elementary reactions and 17 channel species mentioned earlier. The species concentrations for some of the more significant molecules as they vary with distance above the rotating disc are shown in Fig. 2, and it is interesting to observe that more than 50% of the Si deposition, in this case, comes from the flux of SiH_2 to the surface.* The net Si deposition as a function of temperature is shown in Fig. 3. Here we observe the expected result that at low temperatures surface reactions (strongly temperature dependent) govern the deposition rate, and at higher temperatures, the deposition becomes diffusion limited.

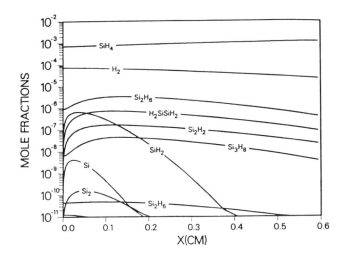

Fig. 2. Species concentrations above a rotating disc reactor for 0.1% silane in helium at atmospheric pressure. Susceptor at 1000°K and a spin rate of 1000 rpm [17].

---

* It must be recognized that any of the results requiring knowledge of the kinetic rate constants, must be considered with caution as these constants are still not well known. For example, new data for the $SiH_4 \rightarrow SiH_2 + H_2$ reaction have recently been reported.[18]

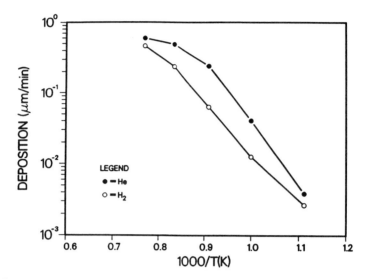

Fig. 3.   Silicon deposition rate versus susceptor temperature for the
same conditions as Fig. 2 [17].

The impinging jet problem has been studied for boron deposition
from BCl_3 and H_2 [19].   Another analysis focuses on the range of
validity of such a simplified one-dimensional analysis when compared
to a proper two-dimensional axisymmetric confined disc [20].   This is
done by assuming a dilute reactant gas, so that the species
concentrations do not affect the flow and temperature distributions.
Then the flow and temperature calculations are compared for the one-
and two-dimensional cases, with bouyancy forces included in the
latter.   Finally, the impinging jet case is used to examine the deposition
of gallium arsenide from organometallic reactants [21].   In this case,
GaAs deposition from either TEG (tri-ethyl-gallium) or TMG (tri-
methyl-gallium) and AsH_3 is studied with 43 chemical species treated
and 51 reactions included.   The composition profiles of some of the
more significant species are shown in Fig. 4, where T is proportional to
distance from the surface.

The authors[21] also observed that for these calculations, within
a temperature range of $850 < T < 1200°K$, that the deposition rate did
not vary with temperature.   Experimental evidence shows that it does
vary.   Therefore, we must keep in mind the fact that the kinetics of
this reaction scheme is so complex and poorly understood that it would
be impractical to expect to be able to predict accurately deposition
rates at this stage.   Further comparison of such calculations with
experimental data will be necessary.

18

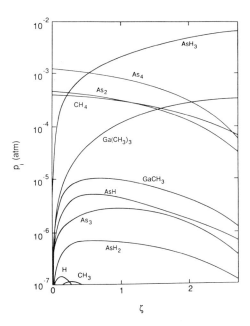

Fig. 4. Species profiles for GaAs deposition from TMG and arsine in H_2 carrier gas. Atmospheric pressure, $p_{TMG} = 3.4 \times 10^{-4}$ atm, $p_{AsH3} = 6.8 \times 10^{-3}$ atm. $T_s = 1000°K$. Ref [21]

Finally, we should note that thermal diffusion as it influences species concentration has been included in calculations of both the impinging jet and the rotating disc [22]. It appears that thermal diffusion significantly (by as much as 20%) <u>reduces</u> deposition rates for silicon from $SiCl_4 + H_2$ in a rotating disc calculation. For the impinging jet case, it <u>increased</u> the deposition rate for boron from $BCl_3 + H_2$ by 7%.[19]

The final example of a one-dimensional flow model is that of a fully developed two-dimensional channel flow, the so-called Poisseulle flow, such as shown in Fig. 5 [23].

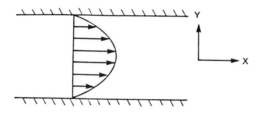

Fig. 5. Channel flow.

The channel is assumed to be doubly infinite in the x direction. There will be an axial pressure gradient, so the pressure will vary along the channel length. We assume the lower wall to be at a constant high temperature, and the upper wall at a constant low temperature. Since mass flows are very small in such systems, variations in density along the channel direction will be small and will be neglected. Similarly, if deposition rates are modest, variations of species concentrations along the channel will also be small. Finally, we consider the lower pressure case, so bouyancy forces are negligible. We also neglect thermal diffusion.

Within these assumptions, we obtain a problem where the unknowns only vary with y, so the problem is one dimensional. The deposition of silicon from $SiH_4 + H_2$ was studied assuming only one gas phase reaction ($SiH_4 \rightleftharpoons SiH_2 + H_2$) and including surface decompositions of SiH_4 and SiH_2. The former was assigned a finite probability, while the latter was assumed to occur with unit probability. The calculated distribution of SiH_2 across the channel is shown in Fig. 6, where we see that the SiH_2 peaks close to the hot wall, as expected. The growth rate of silicon as a function of temperature is shown in Fig. 7. The dependence of growth rate on temperature was not observed to slow down at high temperatures because of the low pressure (300 mTorr) at which the calculations were carried out.

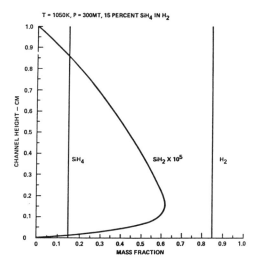

Fig. 6. Species concentrations across channel [23].

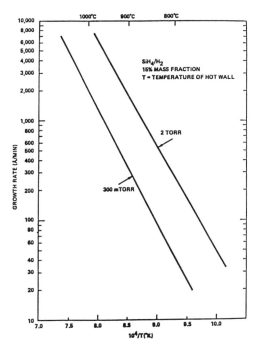

Fig. 7. Deposition rates as a function of temperature [23].

Two-dimensional Flows

Once we abandon the highly simplifying one-dimensional approximation, we face a very difficult numerical task when we attempt to solve the CVD reactor modeling problem with any degree of completeness. If we neglect or highly simplify the chemistry, then unusual two-dimensional geometries and flows can be treated. When we attempt to treat the chemistry accurately and completely, then we are restricted as far as the geometry we can deal with.

Constant Properties Case

To start with, let us consider a general two-dimensional flow with constant properties and uniform temperature. One approach to the numerical solution of this highly nonlinear problem is to replace the governing equations by finite differences and solve the resulting equations by iterative relaxation [24]. Following such a procedure, the author has calculated the flow field for a typical single-wafer, cold-wall CVD reactor. A plot of the streamlines for this case is shown in Fig. 8. As can be seen there, many recirculation zones are formed within the reactor that can contribute to flow uniformity or lack of it. At low pressures, where bouyancy forces are negligible, the principle influence of the high temperature near the wafer will be to increase the gas viscosity in that region of the flow.

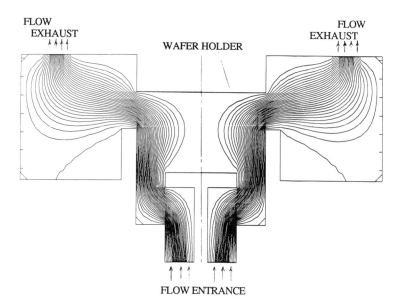

Fig. 8. Streamlines for vertical upflow reactor with wafer held on upper surface facing downward. $R_e = 50$.

Dilute Reactant Case

A major simplification when dealing with the CVD reactor flow problem has been to assume that the reacting gases are heavily diluted with either an inert gas or some product of the reaction (i.e., H_2). For CVD reactors operating in the higher pressure ranges (75 Torr - 1 atmosphere), this is a typical mode of operation (i.e., 1% $SiCl_4$ in H_2), and the approximation is accurate. For low pressure CVD reactors (about 300 mTorr), the reactants are generally less heavily diluted, and this approximation may not be appropriate.

If we can assume that the reactants are heavily diluted, we can estimate the gas viscosity and thermal conductivity as if the gas were entirely the diluent. In other words, $\mu = \mu(p,T)$ and $\lambda = \lambda(p,T)$, and they do not depend on the reactive species concentrations. If at the same time we neglect contributions in the energy equation due to diffusion of enthalpy and heat from net species production, then the momentum and energy equations can be solved independently of the mass species conservation equations. This provides a significant simpliciation of the numerical problem.

One example of the flow and temperature distributions in a rotating disc reaction of <u>finite</u> diameter, including bouyancy effects, is shown in Fig. 9 [17]. The calculations were carried out using an

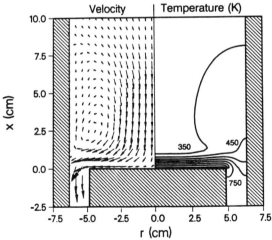

Fig. 9.   He flow and temperatures about a finite rotating disc reactor. $T_s = 1100°K$, $R_e = 1000$, $G_r/R_e^{3/2} = 6.2$, spin rate = 495 rpm [17].

iterative solution of a finite difference approximation to the governing equations. A bouyancy-driven recirculation can be seen, as well as its influence on the temperature distribution. A similar problem has been solved for the impinging jet in a finite reactor chamber [20].

In both of the above papers, an estimate of the validity of the one-dimensional approximation is made by invoking the analogy between heat and mass transfer and comparing calculations of the heat flux (Nusselt number) for both the one- and two-dimensional cases. For the spinning disc case, the influence of spin rate and disc temperature on whether or not the one-dimensional solution is an acceptable approximation to the two-dimensional case is shown in Fig. 10.

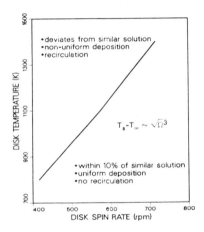

Fig. 10.   Validity of one-dimensional approximation as a function of susceptor temperature and spin rate [17].

Clearly, higher spin rates and lower temperatures lead to situations that can be adequately modeled by a one-dimensional calculation. It should be remembered, however, that no complete calculation of the deposition rate has been reported where the species concentration fields were determined and used to generate the deposition profiles. Therefore, these results should only be taken as qualitative guidelines.

Several additional studies have been done on the confined impinging jet reactor, where MOCVD reactors were of interest [8,25,26]. Since very little, if anything, is known about the kinetics of organometallic reacting gases, these studies assume no gas phase reactions and very fast surface reactions. Such assumptions permit the solution of the mass species equations given the known flow and

24

the susceptor is inverted and the flow is upward [8]. This configuration minimizes bouyancy effects. The numerical solution is obtained using a finite element (Galerkin's method) approach. Here we see that the calculated deposition rates vary with flow rate, as do the measured values, but there is a minimum in the experimental values which is not reproduced by the model.

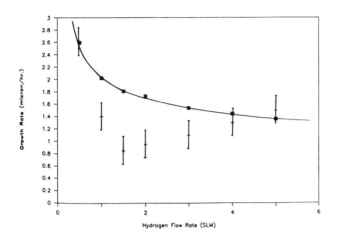

Fig. 11.   Experimental (+) and calculated (•) GaAs growth rate versus H$_2$ flow rate. p = 500 Torr, T$_s$ = 912°K [8].

Fully Coupled Model, Boundary Layer Approximation

A group at Sandia Laboratories has set up a model for the deposition of silicon from SiH$_4$ and H$_2$ or He as diluents in a parallel wall channel similar to Fig. 5 [10,11]. In fact, one could think of this problem as the entrance region to a channel flow where variations along the channel are small. Of course, the entire CVD reactor may be shorter than the entrance region. With this geometric configuration, one can make a boundary layer approximation which converts the elliptic set of model equations to a parabolic set. The parabolic problem is significantly easier to solve numerically than the full elliptic system.

For the high-temperature, atmospheric-pressure reactor, the kinetic scheme is very complex. Initially, these authors considered well over 100 reactions as potentially important. As a result of some numerical sensitivity analysis, these were reduced to only 27. There can be as many as 17 species, so there can be 16 simultaneous species conservation equation in addition to the momentum and energy equations that must be solved simultaneously. The transport properties are taken to depend on the species concentrations as well as p and T. Thermal diffusion can be included in the model, and diffusion of enthalpy and energy due to net production of species are included in the energy equation.

The reaction rates for the homogeneous reactions are partially obtained from the literature and partially estimated by the authors. The probability of decomposition of SiH_4 is taken as $\gamma_{SiH4} = 5.37 \times 10^{-2}$ e^{-9400T}. The assumption is made that the surface reaction probability of all intermediate reactive species is unity, and that for Si_2H_6 and Si_3H_8 are zero.

The results of the numerical solution gave species concentrations throughout the field, and deposition rates were calculated from these [10]. A comparison of calculated rates and experimental data was made for the following case: cell height of 1.25 cm, average flow velocity of 60 cm/sec, 0.76 Torr SiH_4 in 760 Torr (APCVD) or 7.6 Torr (LPCVD) of carrier gas (either H_2 or He), and a location 15-cm (APCVD) or 1-cm (LPCVD) downstream of the entrance. The results are shown in Fig. 12.

Here we observe that the temperature dependence shown by experiment is reproduced by calculation. Also, H_2 as a diluent is shown to suppress deposition, a fact well known from other experimental work. Finally, the fit between experimental results and calculation are unexpectedly good.

In a later paper [11], these same authors modified their model to include thermal diffusion. Again, a specific calculation was reported for 0.7 Torr SiH_4 in 624 Torr of He or H_2, a cell height of 5 cm with a 16-cm/sec average flow velocity and a location 4.5-cm from the susceptor leading edge. The calculation was done both with and without including thermal diffusion, and the results are shown in Table I. For the specific conditions chosen, neglecting thermal diffusion results in an overestimate of the deposition rate of about 50%. Such a correction is certainly significant, and presumably recalculation of the earlier comparison (Fig. 12) will retain the good agreement between calculated and theoretical deposition rates.

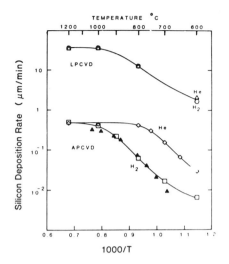

Fig. 12.   Silicon deposition rate versus temperature.   Open symbols are calculations; closed symbols are experimental data [10].

Susceptor temperature (°C)	Carrier gas	Deposition rate[a] with thermal diffusion	Deposition rate[a] without thermal diffusion
550	He	8.8×10^{-4}	1.3×10^{-3}
750	He	1.0×10^{-1}	1.5×10^{-1}
750	H₂	1.1×10^{-2}	1.7×10^{-2}

[a] Deposition rate in μm/min.

TABLE I: Influence of thermal diffusion on silicon deposition rates.[11]

Fully-Coupled Model

One final two-dimensional study has been done of the confined impinging jet model of a CVD reactor, where the same chemistry just described was used but a boundary layer approximation was not appropriate [9].   The primary interest in this study was in the low-pressure, cold-wall CVD reactor as contrasted to the high-pressure MOCVD reactors discussed earlier.   Therefore, gas phase reactions were considered essential and were retained.   At the same time, we also allowed for finite rate reaction probabilities at the surface.

27

Specifically, we chose to consider the decomposition of SiH_4 in H_2 on the hot susceptor. Preliminary calculations including 16 reaction species showed that at the low pressures of interest (300 mTorr) that the reaction $SiH_4 \rightleftharpoons SiH_2 + H_2$ was the primary factor in the deposition of silicon. A calculation done under the following conditions yielded the curves of deposition rate versus susceptor radius, as shown in Figure 13.

Chamber outer diameter:	24.0 cm
Hot plate diameter:	10.0 cm
Distance from gas distributor to hot plate	3.0 cm
Hot plate temperature:	950°K
Chamber and inlet gas temperature:	300°K
Reactant mass flow:	100 sccm
Pressure:	300 mTorr
Mass fractions at inlet:	SiH_4 = 15/30%; H_2 = 85/70%

The SiH_2 concentration normal to the susceptor is shown in Fig. 14 for two radial locations.

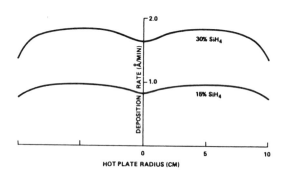

Fig. 13.   Deposition rates versus radius for two concentrations of SiH_4 in H_2 [9].

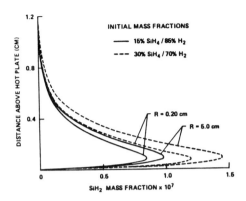

Fig. 14.   SiH$_2$ mass fractions versus distance normal to a hot plate [9].

The SiH$_2$ concentration is close to zero at the hot surface, and rises by orders of magnitude away from the surface. Although the concentrations are small, the contribution to silicon deposition is significant. The rapid change of SiH$_2$ concentration in such a short distance is what causes the numerical difficulties in obtaining solutions to these equations.

Three-dimensional Flows

When bouyancy effects are significant in CVD reactors, the flows can become three-dimensional which further complicates an already complex problem. If we return to the dilute reactant assumption, then the flow and temperature fields can be calculated before tackling the species concentration equations.

The two-dimensional horizontal channel flow examined by Coltrin [11] cannot reproduce the three-dimensional flow phenomena that occurs when the channel has a finite width transverse to the flow. Therefore, two recent studies have examined the finite channel flow that represents the epi reactor used, for example, for either silicon [27] or gallium arsenide [28] thin film deposition at atmospheric pressure. In these studies, the fluid momentum equations were simplified to produce parabolic equations in order to reduce the difficulties of the numerical calculations. For the case with cooled sidewalls, the transverse flow field at three axial locations is shown in Fig. 15, where we can clearly see the longitudinal rolls that arise when the three-dimensional nature of the flow is accounted for.

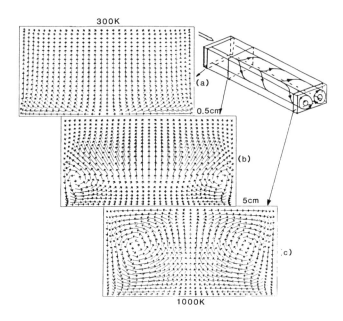

Fig. 15.   Transverse flow velocities in axial flow reactor [27].

Calculations were then made of the species concentrations for TMGa and AsH₃ reactants to deposit GaAs. For this, the deposition rate was calculated and is shown in Fig. 16. These results demonstrate very clearly the nonuniform deposition that can occur in such horizontal epi reactors due to bouyancy effects.

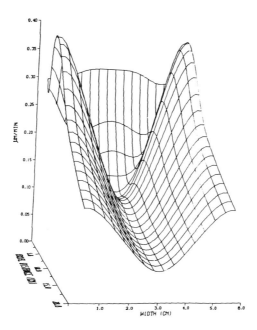

Fig. 16.   Growth rate of GaAs on hot lower wall (susceptor) of axial
flow reactor [28].

In order to model the silicon deposition for the same case, gas phase reactions had to be accounted for. In this case, the authors simplified Coltrin's model by only considering SiH_4, SiH_2, H_2 and Si_2H_6 as the primary growth species. For a case similar to Fig. 15, they obtained deposition profiles as shown in Fig. 17. As can be seen, these are similar to the GaAs result, although the central dip is less severe.

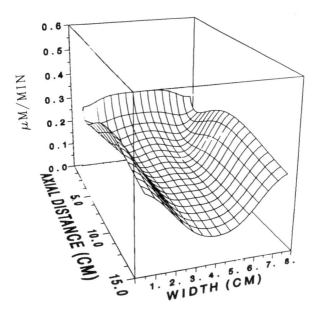

Fig. 17.   Growth rate of Si on hot lower wall (susceptor) of axial flow
reactor [27].

Comparison with Experiment

The traditional method of comparing CVD reactor modeling calculations with experiments uses the deposition rate profiles as the test for validity of the model. In reality, since the chemical kinetics is so poorly understood, what really is needed are measurements in the gas phase as deposition is proceeding. It would be desirable to measure locally the velocities, the temperature and the species concentrations. Since this is obviously a very tall order, it is sufficient initially to measure local temperatures and "selected" species concentrations in order to properly compare experiments with modeling calculations. Such experimental measurements have been reported by the Sandia group [12,13] for the $SiH_4 + H_2$ system. They used pulsed UV laser Raman spectroscopy to measure silane concentrations, laser-excited fluorescence spectroscopy to measure silicon dimer (Si_2) and silicon atom concentrations, and measurements of the intensities of peaks in the rotational Raman spectra of hydrogen to determine local gas temperatures.

The experimental CVD reactor configuration consisted of a heated graphite susceptor (9 x 6.5 x 1.3 cm) placed along the center line of a 10-cm diameter stainless-steel tube with plugs placed below the susceptor to force the gas to flow in a semicircular cross section. Silane diluted with H_2 and/or He was passed over the susceptor at atmospheric as well as lower pressures for a number of temperatures. Quartz windows provided access to the Raman signals. Comparison was made to Coltrin's [11] two-dimensional model, while clearly the experimental flow was three dimensional, so an exact correlation of measured values with calculated values should not be expected.

Temperatures versus height above the susceptor were measured for three susceptor temperatures at atmospheric pressures. The results are shown in Fig. 18. Although the temperatures follow the calculated values quite well, the theoretical values systematically overestimate them. Also, it was not possible to obtain temperature profiles for the low-pressure case (6.2 Torr) with the available experimental technique.

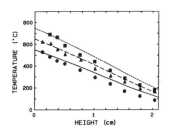

Fig. 18.   Gas temperature versus height above susceptor measured 4-cm from front of susceptor[12].

Concentrations of silane molecules were measured for both atmospheric and low-pressure cases, as shown in Figs. 19 and 20.

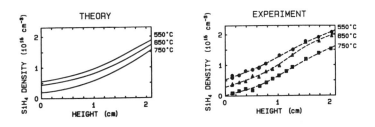

Fig. 19.   Silane density versus height.   Atmospheric pressure [12].

33

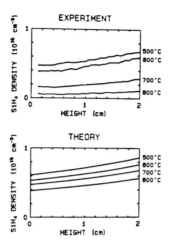

Fig. 20.   Silane density versus height.   Low pressure [12].

Again, comparison is good at atmospheric pressure (although experiment consistently shows more depletion than is calculated), but poor at low pressure.

Qualitative measurements of the silicon dimer (Si_2) concentration above the hot susceptor were made, and are shown in Fig. 21.   The shape of the measured curve is similar to the calculated one, which is clear evidence that the gas phase kinetics is a significant factor in this process.   If the gas phase had been in equilibrium, we would have expected the Si_2 concentration to be highest at the hot susceptor, and for its concentration to fall monatonically from the surface value.

Finally, silicon atom density versus height above the susceptor was measured at atmospheric pressure, and the results are shown in Fig. 22.

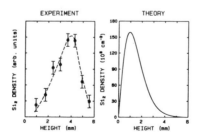

Fig. 21. Si_2 density versus height above susceptor [12].

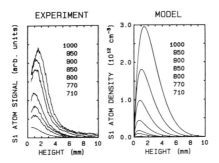

Fig. 22.   Si atom density versus height above susceptor for several
temperatures [13].

Again, the qualitative agreement is good.   The influence of adding H_2 to
a SiH_4 + He reactant is also measured.   The silicon atom concentration is
reduced, as predicted by calculation, but the model predicts a stronger
influence for small additions of H_2 than are seen experimentally.   The
authors feel that the lack of a particle nucleation mechanism in the
theoretical model, when in fact there is some particle nucleation in the
SiH_4 + He reactant case, explains this effect.   In pure SiH_4 + He, if
nucleation occurred, then the experimentally measured values of
silicon atom concentration should be less than the model predictions, as
was seen.   The remaining measurements looked at the influence of
pressure.   Again, agreement was good for atmospheric pressure
experiments and poor for low-pressure ones.   The explanation here is
that the 3-dimensional versus 2-dimensional discrepancies become
significant at low pressures and cause the errors.

Conclusions

In the past, many attempts were made to model CVD reactors in
order to provide a vehicle for the design of such systems.   Because of
the complexity of the nonlinear Navier-Stokes equations governing the
flow and temperature fields, and the lack of information on the
kinetics both in the gas phase and the surface, such models were
inadequate at best.   More recently, high-speed digital computers have
made flow and temperature solutions feasible, and some kinetic data
has been developed at least for the SiH_4 + H_2 reaction system.
Accordingly, more realistic models are now being developed from "first
principles", and there is hope for their practical utility.

In this paper, we have reviewed recent modeling studies for one, two- and even three-dimensional CVD reactors and have attempted to define their limitations. We have also described how, for the first time, experimental details of the gas phase have been measured and compared to modeling calculations.

Although much progress has been made recently, it is also clear that truly practical development of such techniques will require a great deal of chemical kinetic data about a number of reactions of interest. The incentive to develop such data will increase as modeling techniques become more sophisticated.

References

1. G. W. Sutton and A. Sherman, Engineering Magnetohydrodynamics (McGraw-Hill, New York, 1965).

2. V. S. Ban, J. Crys. Growth 45 (1978), 97.

3. C. W. Manke and L. F. Donaghey, J. Electrochem. Soc. 124 (1977), 561.

4. J. Juza and J. Cermak, J. Electrochem. Soc. 129 (1982), 1627.

5. K. F. Roenigk and K. F. Jensen, J. Electrochem. Soc. 132 (1985), 448.

6. C. Houtman, D. B. Graves and K. F. Jensen, J. Electrochem. Soc. 133 (1986), 961.

7. H. Moffat and K. F. Jensen, J. Crys. Growth 77 (1986), 108.

8. P. Lee, D. McKenna, D. Kapur and K. F. Jensen, J. Crys. Growth 77 (1986), 120.

9. A. Sherman, Proceedings of the Symposium on Reduced Temperature Processing for VLSI, Vol. 86-5 (Electrochem. Soc., New Jersey, 1986).

10. M. E. Coltrin, R. J. Kee and J. A. Miller, J. Electrochem. Soc. 131 (1984), 425.

11. M. E. Coltrin, R. J. Kee and J. A. Miller, J. Electrochem. Soc. 133 (1986), 1206.

12. W. G. Breiland, M. E. Coltrin and P. Ho, J. Appl. Phys. 59 (1986), 3267.

13. W. G. Breiland, P. Ho and M. E. Coltrin, J. Appl. Phys. 60 (1986), 1505.

14. H. Liepmann and A. Roshko, Elements of Gasdynamics (John Wiley & Sons, New York, 1957).

15. H. Schlichting, Boundary Layer Theory (McGraw-Hill, New York, 1979).

16. R. Pollard and J. Newman, J. Electrochem. Soc. 127 (1980), 744.

17. R. J. Kee, G. H. Evans and M. E. Coltrin, Sandia National Laboratories Report SAND87-8660 (1987).

18. J. M. Jasinski, J. Phys. Chem. 90, 555 (1986).

19 M. Michaelidis and R. Pollard, J. Electrochem. Soc. 131 (1984), 860.

20 C. Houtman, D. B. Graves and K. F. Jensen, J. Electrochem. Soc. 133 (1986), 961.

21. M. Tirtowidjojo and R. Pollard, Proceedings of the First International Conference on Processing of Electronic Materials (Engineering Foundation, New York, 1987).

22. J. P. Jenkinson and R. Pollard, J. Electrochem. Soc. 131 (1984), 2911.

23. A. Sherman, Chemical Vapor Deposition (Noyes Publishing, New Jersey, 1987).

24. O. R. Burggraf, J. Fluid Mech. 24 (1966), 113.

25. D. I. Fotiadis, A. M. Kremer, D. R. McKenna and K. F. Jensen, J. Crys. Growth 85 (1987), 154.

26. Y. Kusumoto, T. Hayashi and S. Komiya, Japn. J. Appl. Phys. 24 (1985), 620.

27. K. F. Jensen, H. Moffat and K. F. Roenigk, Proceedings of the First International Conference on Processing of Electronic Materials (Engineering Foundation, New York, 1987).

28. H. Moffat and K. F. Jensen, J. Crys. Growth 77 (1986), 108.

Microstructure of Reactively Sputtered Oxide Diffusion Barriers

E. Kolawa, C. W. Nieh, F. C. T. So[*] and M-A. Nicolet

California Institute of Technology, Pasadena, CA 91125
Hewlett-Packard Co., San Jose, CA 95131.

Abstract

Molybdenum oxide ($Mo_{1-x}O_x$) and ruthenium oxide (RuO_2) films were prepared by rf reactive sputtering of Mo or Ru targets in an O_2/Ar plasma. Both films exhibit metallic conductivities. The influence of the deposition parameters on the phase that forms and on the microstructure of $Mo_{1-x}O_x$ and RuO_2 films is reported. A phase transformation is observed in $Mo_{1-x}O_x$ films subjected to heat treatment. The diffusion barrier performance of $Mo_{1-x}O_x$ and RuO_2 layers interposed between Al and Si is compared.

Microstructural Science for Thin Film
Metallizations in Electronics Applications
Edited by J. Sanchez, D.A. Smith and N. DeLanerolle
The Minerals, Metals & Materials Society, 1988

Introduction

An electrical contact to a semiconductor must satisfy a prescribed electrical characteristic and this characteristic must be stable with time. In silicon integrated circuits, aluminum is commonly used for contacts and interconnections, but the high solubility and diffusivity of silicon in aluminum causes a degradation of contacts. To minimize such deleterious interactions, diffussion barriers between Al and Si have been introduced in contact metallizations (1-3). The ideal barrier layer should be thermodynamically stable with Al and Si (or silicide), constitute a kinetic barrier to prevent the transport of Al or Si across the barrier, be electrically conducting and adhere well to both Si (or silicide) and the Al top layer. An ideal barrier does not exist (2). A real barrier always fails in a particular metallization in one way or the other when one or more of the above conditions are not satisfied.

Various interstitial alloys like nitrides (4-5), borides (6) and carbides (7) have been investigated in the past as diffusion barriers in contact metallizations. No attention was paid to thin films of transition metal oxides as barrier layers. There are, however, transition metal oxides that exibit metallic conductivities at room temperature. Some of these oxides have an oxygen to metal ratio of 2. One group consists of the dioxides of RuO_2, OsO_2, IrO_2 and RhO_2 which crystallize in rutile structure. The other group consists of the dioxides which adopt distorted variants of the rutile structure: CrO_2, MoO_2 and WO_2. Recently, RuO_2 deposited by MOCVD (8) or reactive sputtering (9-12) and $Mo_{1-x}O_x$ deposited by reactive sputtering (13-14) were reported to be most effective diffusion barriers between Al and Si. RuO_2 and MoO_2 are dissimilar in their stability. RuO_2 is the only thermodynamically stable oxide of Ru at room temperature (15). On the other hand, there exist higher oxides of Mo that are electrically insulating (MoO_3). In this paper we compare sputtering processes, microstructure and diffusion barrier properties of RuO_2 and $Mo_{1-x}O_x$ films.

Experimental Procedure

Silicon <111> wafers,"<Si>", silicon wafers covered by thermally grown SiO_2 (3000A), carbon and NaCl were used as substrates for the films. Prior to loading into the deposition chamber, the silicon wafers were passed through organic cleaning steps in ultrasonic baths of TCE, acetone and methanol and were then etched with diluted HF. All films of this study were deposited by reactive rf sputtering using a 7.5 cm diameter planar magnetron cathode. The substrate holder was placed about 7 cm below the target and was neither cooled nor heated externally. A base pressure of 1×10^{-6} Torr in the sputtering chamber was attained prior to $Mo_{1-x}O_x$ or RuO_2 deposition in O_2/Ar ambients. The total gas pressure of the premixed O_2/Ar was adjusted with a variable leak valve and monitored with a capacitive manometer prior to striking the discharge (referred to as "initial" total gas pressure). Both $Mo_{1-x}O_x$ (or Mo) and RuO_2 (or Ru) films described in this paper were sputtered with an initial total gas pressure of 10 mTorr and zero substrate bias. The relative initial partial pressure of oxygen in Ar, defined as the ratio of the initial partial pressure of oxygen to the initial total gas pressure $p(O_2)/p(O_2+Ar)$ was varied from 0% to 60%. The thickness of the films was in the 80-200 nm range. The composition of films was measured by backscattering spectrometry on films deposited on carbon substrates. The film resistivities were determined from sheet resistivities obtained from four point probe data and thicknesses were measured with a profilometer for films on oxidized Si substrates.

Transmission electron microscopy and x-ray Read camera diffraction were used to study the phases and microstructure of the films. Plan-view TEM specimens of the films were prepared by chemical thinning from the back side of the silicon substrates or by depositing the thin films (30 nm) onto NaCl crystals and then transferring them to a copper grid. <Si>/$Mo_{80}O_{20}$/Al and <Si>/RuO_2/Al samples were prepared to test the diffusion barrier properties of RuO_2 and $Mo_{1-x}O_x$ films. The RuO_2 films (40 nm) or $Mo_{80}O_{20}$ films (70-100 nm) and Al overlayers (300-500 nm) were sputter-deposited sequentially onto the Si substrates without breaking vacuum.

The effectiveness of these barriers was evaluated by electrical measurements performed on shallow As^+ implanted n+/p diodes with <Si>$Mo_{80}O_{20}$/Al or <Si>/RuO_2/Al contact structures, by backscattering spectrometry and by cross-sectional TEM analysis. Details of the fabrication procedure of the Si n^+/p diodes (0.35 μm junction depth, 500x500 $μm^2$ junction area, 300x300 m^2 contact window area), are reported elsewhere (16). To prepare cross-sectional TEM of the contact structures the samples were first glued together face to face, followed by mechanical thinning to 10 m. Finally argon ion milling at liquid nitrogen temperature was used to thin the specimen to electron transparency. Annealing of samples was carried out in a vacuum of better than $1x10^{-6}$ Torr in the temperature range of 350°-700°C for different annealing durations.

Results and Discussions

1) Deposition

Figure 1 presents the oxygen concentration in the $Mo_{1-x}O_x$ and the deposition rate of films as a function of forward sputtering power. Films were deposited with three different initial partial pressures of oxygen. For a fixed sputtering power, an increase in the inital partial pressure of oxygen raises the oxygen content in the films. When the oxygen partial pressure is fixed, the oxygen content in the films decreases continuously as the sputtering power is raised from 100W to 600W. The deposition rate of $Mo_{1-x}O_x$ films is proportional to the forward sputtering power and practically independent of the initial partial pressure of oxygen.

Figure 2 shows the dependence of growth rate of RuO_2 or Ru films on the forward sputtering power and on the initial oxygen partial pressure. Stochiometric RuO_2 films are obtained only with low forward powers (100 or 200W) in combination with initial partial pressures equal and higher than 10% and 20% for 100W and 200W, respectively.

By comparing the sputtering processes of these two materials it is clear that the deposition behavior of RuO_2 contrasts that of $Mo_{1-x}O_x$. The amount of oxygen incorporated into the ruthenium-oxygen films is discontinous: for given deposition conditions either Ru or RuO_2 is deposited. On the other hand, the oxygen concentration in the $Mo_{1-x}O_x$ films can be continuously varied for the entire range of sputtering parameters. This dissimilarity suggests a different formation mechanism of $Mo_{1-x}O_x$ and RuO_2 films.

2) Phases and Microstructure

Figure 3 shows the microstructure and diffraction pattern of the $Mo_{50}O_{50}$ and $Mo_{60}O_{40}$ films obtained by reactive sputtering with the 40% and 30% oxygen partial pressure, respectively. The forward sputtering power was the same (200W) for both films. A single fcc Mo-O phase is detected in both films by x-ray and electron diffraction. There is no evidence of the

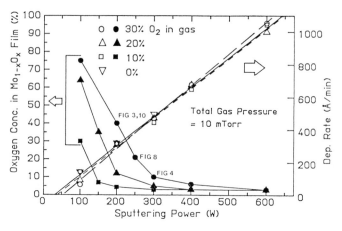

Figure 1 - Oxygen concentration in $Mo_{1-x}O_x$ films and depostion rates for samples sputtered in 10, 20, and 30% of oxygen in the sputtering gas. The initial total gas pressure is 10 mTorr and the substrate bias is zero.

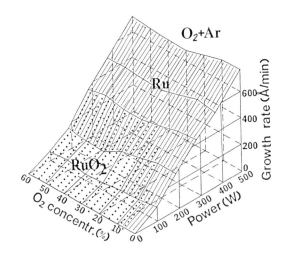

Figure 2 - The growth rate of RuO_2 (Ru) films as functions of the forward sputtering power and the initial partial pressure of oxygen in the O_2/Ar gas. The initial total gas pressure is 10 mTorr and the substrate bias is zero. The dotted and lined areas identify sputtering conditions that lead RuO_2 and Ru film formation, respectively. The overlays of these areas indicates a transition from Ru to RuO_2 whose exact position and nature are not known.

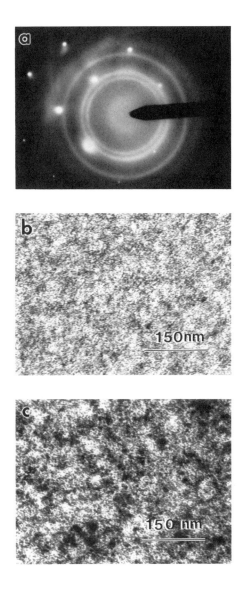

Figure 3 - Diffraction pattern (a), and
bright-field micrograph (b) of $Mo_{60}O_{40}$ film;
bright-field micrograph (c) of $MO_{50}O_{50}$ film
(forward sputtering power: 200W; substrate bias:
OV; total initial pressure: 10 mTorr).

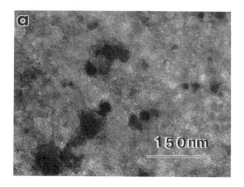

Figure 4 - Bright-field micrograph (a) and diffraction pattern (b) of $Mo_{90}O_{10}$ films (forward sputtering power: 300W, substrate bias: 0V, total initial pressure: 10 mTorr with 30% O_2).

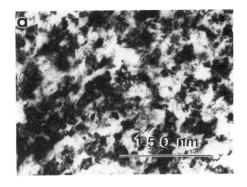

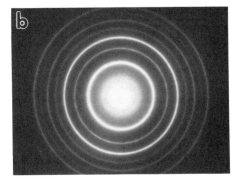

Figure 5 - The bright-field micrograph (a) and
diffraction pattern (b) of Mo film (forward
sputtering power: 300W; Substrate bias: OV; total
initial presssure: 10m Torr with 0% O_2).

MoO_2 or MoO_3 phases in spite of the significant amount of oxygen incorporated to the films. The average grain size of the $Mo_{50}O_{50}$ film is slightly larger (about 20nm) than the average grain size of the $Mo_{60}O_{40}$ sample (about 10nm).

Two $Mo_{1-x}O_x$ films were deposited with the same (30%) initial partial pressure of oxygen and two different forward powers. The composition of the films is $Mo_{60}O_{40}$ (for 200W) and $Mo_{80}O_{20}$ (250W). A single fcc Mo-O phase was detected in both films by electron diffraction. The sputtering power clearly influences the grain size of samples because the average grain size of samples sputtered with the low power (200W) are significantly smaller (10nm) than the average grain size of the film sputtered with high power (40nm). The microstructure of the $Mo_{60}O_{40}$ film is presented in Fig. 3 and the microstructure of the $Mo_{80}O_{20}$ film is presented in Fig. 8 later in this paper.

The microstructures and diffraction patterns of the $Mo_{90}O_{10}$ and Mo films deposited with the 300W forward power, the 30% and 0% initial oxygen partial pressure, are presented in Figures 4 and 5, respectively. The $Mo_{90}O_{10}$ film is observed to contain bcc Mo grains embedded in a background material that appears amorphous under TEM. The average size of the bcc Mo grains is about 50 nm. The pure molybdenum film is of a single phase with bcc structure and has grains of about 50 nm size. According to Chopra (17) the normal bcc phase of Mo is obtained in sputter deposited films only when the substrate temperature exceeds 400°C; while the fcc phase of Mo is obtained in the 200-400°C range, and amorphous Mo films are obtained below that temperature range. It is also reported there that the formation of different phases depends on sputtering parameters (deposition rate, deposition technique, substrate temperature etc.) and is not connected with a presence of impurities in films. We find that the presence of the bcc or fcc phase is determined by the amount of oxygen incorporated into films. Films with low oxygen content (<10at.%) are of the normal bcc Mo phase, whereas films with high concentration of oxygen (>10at.%) are fcc. Only the $Mo_{25}O_{75}$ film is composed of a single MoO_3 phase, as determined by x-ray diffraction. We do not observe any oxide phases in $Mo_{1-x}O_x$ for 0<x<60. In these films, the oxygen atoms are probably incorporated in the Mo grains as interstitials, and decorate the grain boundaries as well. The lattice constant of fcc Mo in our films is indeed 2% larger than that reported by Chopra. Interstitial oxygen atoms evidently stabilize the metastable fcc Mo structure.

Figure 6 shows the microstructure and the diffraction pattern of RuO_2 samples sputtered with oxygen initial partial pressures of 30%,(a and b) 40%, (c) and 60% (d) at a constant forward power of 200W. The films consist of fine equiaxed grains and the structure is uniform. The slight increase in grain size from 8 nm to 14 nm is observed as the partial pressure of oxygen increases from 30% to 60%.

3) Resistivity

The resistivities of the RuO_2 films are in the range of 150-200 cm. (The bulk resistivity is $46 \mu\Omega$ cm^2 (18)). An annealing of these samples in vacuum significantly lowers the resistivity. For example, after 15 min, at 800°C for the resistivity is about one third of its initial value. This decrease in resistivity is associated with a growth of the crystalline grains of RuO_2. No phase tranformation was observed.

All as-deposited $Mo_{1-x}O_x$ films are electrically conducting except $Mo_{25}O_{75}$ which corresponds to the insulating MoO_3 phase. The resistivity of as deposited $Mo_{1-x}O_x$ films increases monotically with the oxygen content in the films (Fig. 7) regardless of the sputtering conditions. A similar trend has also been observed for reactively sputtered W-N films (19).

46

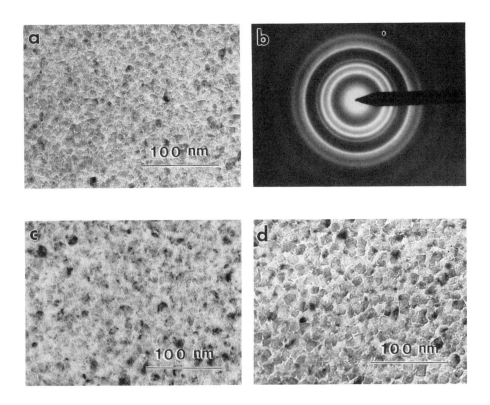

Figure 6 - Bright-field micrograph (a) and
diffraction pattern (b), of RuO_2 film sputtered
with 30% oxygen initial partial pressure;
bright-field micrographs of RuO_2 films sputtered
with 40% (c) and 60% (d) oxygen initial partial
pressure (forward sputtering power: 200W; substrate
bias: OV; total initial pressure: 10 mTorr).

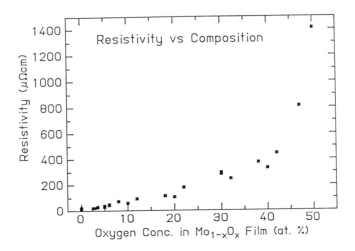

Figure 7 - Resistivity of $Mo_{1-x}O_x$ films as a
function of their oxygen concentration (for all
forward sputtering powers and oxygen initial
partial pressures; substrate bias: Ov; total
initial pressure: 10 mTorr).

4) Annealing of Films

The microstructure and diffraction patterns of the as-deposited and
annealed (600° and 800°C for 30 min) $Mo_{80}O_{20}$ samples are presented in
Figure 8 and 9, respectively. Electron diffraction patterns establish that
a phase transformation from the metastable fcc Mo to the bcc Mo has taken
place after annealing at 600°C for 30 min. The average grain size of the
new bcc phase formed at 600°C is smaller (9 nm) than the as-deposited
sample (40 nm). The electron diffraction pattern of the $Mo_{80}O_{20}$ sample
annealed at 800°C for 30 min shows (Figure 9) the presence of the bcc Mo
and MoO_2 phases. The average grain size of the new bcc phase formed at
800°C increased to 100nm, Figure 8c. The annealing behaviour of the
$Mo_{60}O_{40}$ sample is very similar to that of $Mo_{80}O_{20}$ except that a mixture of
bcc Mo and MoO_2 phases already exists after annealing at 600°C for 30 min
(Fig. 10). The resistivities of the $Mo_{1-x}O_x$ films are unaffected by
annealing at 600°C, but drop abruptly for heat treatments between 600°C and
800°C. The resistivities of $Mo_{80}O_{20}$ and $Mo_{60}O_{40}$ films annealed at 800°C
are 10 and 25 μΩ cm, respectively. These values are comparable to the
resistivity of the pure Mo film annealed at 800°C (10 cm). Backscattering
analysis detects no oxygen loss from the $Mo_{1-x}O_x$ films after any of the
annealing treatments described here.

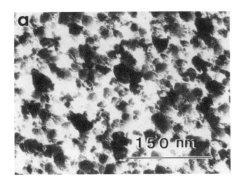

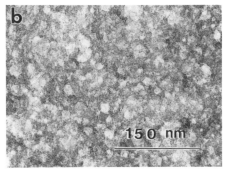

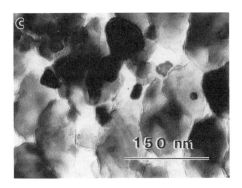

Figure 8 - Bright-field micrograph of $Mo_{80}O_{20}$ sample: as-deposited (a), annealed at 600°C for 30 min (b), and annealed at 800°C for 30 min (c).

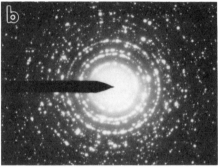

Figure 9 – Diffraction pattern of the $Mo_{80}O_{20}$ films shown in Fig. 8: as –deposited (a) and annealed at 800°C for 30 min (b).

5) Diffusion Barrier Tests

The RuO_2 and $Mo_{1-x}O_x$ films were tested as diffusion barriers between Al and Si. To determine the interaction between the layers, backscattering spectrometry, electrical measurements and XTEM were used. Backscattering analysis of $<Si>/RuO_2/Al$ samples shows that Al-Si interdiffusion can be suppressed by a RuO_2 barrier of 40 nm thickness up to 600°C for 30 min. A $<Si>/TiSi_2/RuO_2/Al$ metallization was tested on n^+/p shallow junction diodes of $350\,\mu$ m junction depth. The leakage current of the diodes remains unaltered after annealing at 600°C for 30 min (see Fig. 11 (from ref. 9)). The diodes subjected to heat treatment at 650°C for 30 min were shorted. Electrical measurements reported previously (20) for shallow p/n junctions with the $<Si>/TiSi_2/Al$ metallization, i.e. without a RuO_2 diffusion barrier, showed that the diodes were shorted after annealing at 400°C for 30 min. It is clear that the RuO_2 film enhances the thermal stability of the $<Si>/TiSi_2/Al$ contact. The RuO_2 film raises the failure temperature by 200°C.

The cross-sectional structure of an as-deposited $<Si>/RuO_2/Al$ sample is shown in Fig. 12. A thin and laterally uniform amorphous layer is present at the $<Si>/RuO_2$ interface. This layer has all the TEM features characteristic of SiO_2, and we identify it with SiO_2. Figure 13 shows the same structure after annealing at 650°C for 60 min. The layered structure of the sample is preserved showing that the RuO_2 film prevents extensive interaction between the Si and Al. However, a new crystalline interfacial

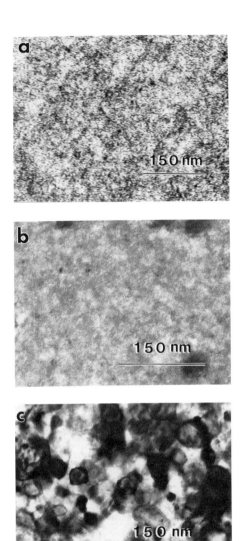

Figure 10 - Bright-field micrograph of $Mo_{60}O_{40}$ sample: as-deposited (a), annealed at 600°C for 30 min (b), and annealed at 800°C for 30 min (c).

layer at the RuO_2/Al interface appears that has an average thickness of about 10 nm. This layer starts to develop at 450°C after annealing for 6 hrs. Attempts were made to identify this compound but its diffraction pattern cannot be attributed to any known Al-Ru compound or any known Al_2O_3 structure. The formation of Al_2O_3 would be thermodynamically favored because the heat of formation of Al_2O_3 (-446 g-cal/mol) is much more negative than that of RuO_2 (-57.3 g-cal/mol). Since Al_2O_3 exists in many different phases (depending on the formation conditions), the compound we observe may be a phase of Al_2O_3 that has not been reported previously. We can not exclude the possibility that this interfacial layer is a ternary Al-Ru-O compound.

These two interfacial layers on either side of the RuO_2 film, SiO_2 and Al_2O_3 (or Al-Ru-O ternary), are quite probably the reason for the excellent performance of RuO_2 as a diffusion barrier. These interfacial layers are thin enough to permit the flow of current through them by electron tunneling. The interfacial layers may also be self-sealing, which would impart a regenerative property to this contact structure. This idea is further supported independently by the fact that $Mo_{1-x}O_x$ films also act as excellent diffusion barriers between Al and Si.

Figure 14 shows the histograms of the reverse currents of 40 n^+p diodes with $<Si>/Mo_{80}O_{20}/Al$ contacts measured before and after annealing at 600°C for 20 min. No junction shorting is observed. Backscattering spectra show a small outdiffusion of Mo into the Al after annealing at 600°C for 40 min. Electrical measurements showed that this outdiffusion does not degrade the junction characteristics. Our experiments (13) with other $Mo_{1-x}O_x$ barriers showed that the film must contain at least 15at.% of oxygen to act as a good barrier between Al and Si at 600°C.

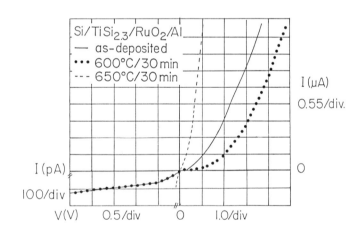

Figure 11.- Electrical characteristics (I-V) of shallow n^+/p diode with a $<Si>/TiSi_2/RuO_2/Al$ metallization before and after annealing at 600°C and 650°C for 30 min.

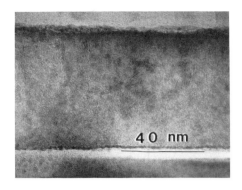

Figure 12 - XTEM micrograph of an as-deposited
<Si>/RuO$_2$(40 nm)/Al(20 nm). The Si<111> substrate
is at the bottom; the Al film is at the top.

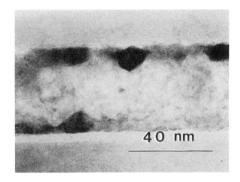

Figure 13 - XTEM micrograph of <Si>/RuO$_2$/Al sample
(as shown in Fig. 12) annealed in vacuum at 650°C
for 1 hr.

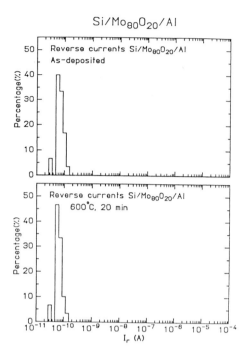

Figure 14 - Histograms of reverse diode leakage currents for <Si>/Mo$_{80}$O$_{20}$(100 nm)/Al(500 nm) contacts before and after annealing in vacuum at 600°C for 20 min.

Conclusions

We have investigated the properties and microstructure of reactively sputtered RuO$_2$ and Mo$_{1-x}$O$_x$. The features of the rf sputter deposition process and the resulting film characteristics are quite different for the two cases. Yet, the diffusion barrier performances of these two oxides are equally outstanding and superior to any previously investigated barrier layers. We attribute this fact to the presence of oxygen in the barrier which promotes the in-situ formation of interfacial oxide layers with Si and Al that are electronically permeable and self-sealing.

Acknowledgements

We would like to thank Rose Pieters-Emerick for her help with manuscript preparation and Rob Gorris for technical assistance. We acknowledge the financial support of the National Science Foundation under MRG Grant DMR-8421119. We also thank Intel Corporation for a grant.

References

1. M-A. Nicolet, Thin Solid Films, 54, 415 (1978).

2. M-A. Nicolet and M. Bartur, J. Vac. Sci. Technol. 19, 786 (1981).

3. M. Wittmer, J. Appl. Phys. 53, 1007 (1982).

4. S. Kanamori, Thin Solid Films, 136, 195 (1985).

5. H. P Kattelus, E. Kolawa, K. Affolter and M-A. Nicolet, J. Vac. Sci. Technol. A 3, 2246 (1985).

6. J. R. Shappirio, Y. Y. Finnegan, R. A. Lux and D. C. Fox, Thin Solid Films, 119, 23 (1984).

7. M. Eizenberg, S. P. Muranka and P. Heinemann, J. Appl. Phys. 54, 3195 (1983).

8. M. L. Green, M. E. Gross, L. E. Papa, K. Y. Schnoes and D. Brasen, J. Electrochem. Soc. 132, 2077 (1985).

9. E. Kolawa, F. C. T. So, E. T-S. Pan and M-A. Nicolet, Appl. Phys. Lett. 50, 854 (1987).

10. L. Krusin-Elbaum, M. Wittmer and D. S. Yee, Appl. Phys. Lett. 50 1879 (1987).

11. E. Kolawa, F. C. T. So, W. Flick, X.-A. Zhao, E. T-S. Pan and M-A. Nicolet, submitted to Thin Solid Films.

12. C. W. Nieh, E. Kolawa, F. C. T. So, and M-A. Nicolet, submitted to Mat. Lett.

13. F. C. T. So, E. Kolawa, X.-A. Zhao, E. T-S. Pan and M-A. Nicolet, Appl. Phys. A (in press).

14. F.C.T. So, E. Kolawa, X.-A. Zhao, T.-S. Pan, M-A. Nicolet, Journal Vac. Sci. & Tech. B 5, 1748, (1987).

15. F. A. Shunk, Constitution of Binary Alloys, Second Supplement (McGraw-Hill, New York, 1965).

16. F. C. T. So, X.-A. Zhao, E. Kolawa, J. L. Tandon, M. F. Zhu and M-A. Nicolet, Mat. Res. Soc. Symp. Proc. Vol. 54, ed. R. J. Nemanich, P. S. Ho and S. S. Lau (MRS, Pittsburgh, 1986), p. 139.

17. K. L. Chopra, M. R. Randlett and R. H. Duff, Phil. Mag. 16, 261 (1967).

18. W. D. Ryder and A. W. Lawson, Phys. Rev. B1, 1494 (1970).

19. E. Kolawa, F. C. T. So, X.-A. Zhao and M-A. Nicolet, in Tungsten and Other Refractory Metals for VLSI Applications II, edited by E. K. Broadbent (MRS, Pittsburg, 1987), p. 311.

20. C. Y. Ting and M. Wittmer, J. Appl. Phys. 54, 937 (1983).

INTERFACIAL REACTIONS OF

COBALT THIN FILMS ON (001) GaAs

F.Y. Shiau and Y.A. Chang
Department of Metallurgical and Mineral Engineering
University of Wisconsin-Madison
Madison, WI 53706

L.J. Chen
Department of Materials Science and Engineering
National Tsing Hua University
Hsinchu, Taiwan, R. O. C.

Abstract

Interfacial reactions between cobalt thin films and (001) GaAs have been studied by transmission electron microscopy , energy-dispersive analysis of X-rays in a scanning TEM, Auger electron spectroscopy and X-ray photoelectron spectroscopy. The completely reacted layer was found to be " β-Ga_2O_3/ (CoGa, CoAs) /GaAs" . The formation of a surface layer of β-Ga_2O_3 and the use of encapsulated samples minimized As loss from the reacted layer. Both CoGa and CoAs were found to grow epitaxially on (001) GaAs. The orientation relationships between CoGa and GaAs were determined to be [001] CoGa // [001] GaAs and (220) CoGa // (220) GaAs. The Burgers vectors of interfacial dislocations were identified as 1/2 <101> and 1/2 <011> which are inclined to the (001) GaAs surface. Almost all of the CoGa films were found to be epitaxially related to the surface. No interfacial dislocations were observed in most of the epitaxial CoAs films which are considered to be pseudomorphic with respect to GaAs. The orientation relationships between CoAs and GaAs were determined to be [101] CoAs // [001] GaAs and (020) CoAs // (220) GaAs. Two-step annealing was found to be effective in promoting epitaxial growth.

Microstructural Science for Thin Film
Metallizations in Electronics Applications
Edited by J. Sanchez, D.A. Smith and N. DeLanerolle
The Minerals, Metals & Materials Society, 1988

Introduction

One of the essential features in designing microelectronic devices is the need for connections between active elements and thus for active contacts. The need for reliable, low-resistance, reproducible, and stable ohmic contacts and Schottky diodes to III-V compound semiconductors such as GaAs is particularly great [1,2]. Since the electrical properties of metal - GaAs (M/GaAs) contacts are sensitive to the interfacial regions, a basic understanding of the structure, chemistry and morphology at the interfaces is essential in the designing of a new contact metallization scheme. Most of the GaAs contact schemes available today do not have adequate electrical uniformity, thermal stability, reliability and reproducibility to meet the stringent requirements for high-speed and optoelectronic applications.[3] It is therefore a great challenge to fabricate stable metal contacts with precise lateral uniformity, penetration depth and desired electrical properties in the integration and miniaturization of high performance GaAs devices.

Alloyed metallization schemes such as Au-Ge-Ni/GaAs[4,5] have been extensively studied for ohmic contacts. Au is known to act as a base metal and Ge acts as a doping element. This system typically exhibits complex phase formation, interface morphology and growth kinetics. It is of particular interest to elucidate the role of the near-noble metal films and to exploit the potential of these films as contact materials to GaAs.

Compared to the rather extensively studied systems such as Pt/GaAs[6-9], Pd/GaAs[8-16] and Ni/GaAs[8,9,17-20], few studies have been carried out on Co/GaAs. Previous investigations of Co/GaAs reactions[21] showed that highly oriented Co_2GaAs was initially observed at 325- 500 °C. However, the crystal structure and the crystallographic relationships of Co_2GaAs with respect to (001) GaAs are as yet undetermined. Subsequent annealing resulted in the formation of the binary phases CoGa and CoAs. Sands et al.[22] reported that both CoGa and CoAs were preferentially oriented with respect to the (001) GaAs substrate after annealing 20 min at 600 °C. Genut and Eizenberg[23] showed that the reaction is governed by diffusion of Co toward GaAs. Palmstrom et al.[24] demonstrated that the reaction depends on the temperature. According to them, when the annealing temperature was increased, the inward diffusion of Co was surpassed by the outward diffusion of Ga and As, which in turn induced the formation of binary phases. In this paper, the phases formed on the Co/GaAs interface and the morphological relationships between these phases with respect to GaAs were characterized by TEM, TED and XTEM. Auger electron spectroscopy (AES) with Ar^+ ion sputtering was performed to investigate the vertical elemental redistributions. Lateral redistributions were examined using the energy dispersive X-ray analyzer in the scanning TEM. X-ray photoelectron spectroscopy combined with Ar^+ ion etching was also used to characterize the reacted layer.

Experimental Procedures

Undoped semi-insulating (ρ >$10^7 \Omega$-cm) (001) GaAs wafers were degreased in acetone and trichloroethylene (TCE), etched in 50 % HCl solution for 2 min, rinsed in deionized H_2O, and then dried with N_2 gas before being loaded into the evaporation chamber. Cobalt films 40 nm thick were deposited onto the wafers by electron beam evaporation under a vacuum of 10^{-7} Torr. The samples were encapsulated in 5 mm ID quartz tubes under a vacuum of 10^{-4} Torr for subsequent annealing. The encapsulated samples were annealed at various temperatures between 250 - 800 °C for 1 h. and quenched into water. A two-step annealing process was used to promote the epitaxial growth of Co-compounds on GaAs. The first step was to anneal at 250 °C for 1 h. followed by a subsequent anneal at 750 °C for 1 h. Some of the samples were heat treated at 300 °C for 1-45 h. to study the initial phase formed.

Plan-view TEM specimens were prepared by back-etching from the GaAs side with a 5 % bromine-methanol solution. Cross-sectional specimens were prepared by Ar⁺ ion milling. The thin foils were then examined using a JEOL 100 B TEM and a JEOL 200 CX scanning transmission electron microscope equipped with an energy-dispersive X-ray analyzer (EDS).

Composition depth profiles were obtained using a PHI-548 Auger spectrometer in the dN(E)/dE broadscan mode. Sputtering was performed using a Perkin-Elmer differential ion gun with 3 KeV argon ions , a 25 micro-amp ion current and a raster size of approximately 4x4 mm². The background pressure of the Auger system during sputtering was 2 x 10⁻⁷ Torr. A PHI-5400 ESCA system equipped with Ar⁺ ion sputtering was used to study the reacted layer. The sputtering was performed with 4 KeV Ar⁺ ions under a background pressure of 5 x 10⁻⁸ Torr.

Results

The as-deposited Co films were determined to be amorphous when analyzed by TEM in

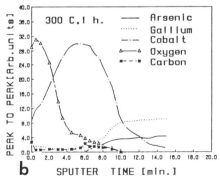

Fig. 1. (a) TEM micrograph and diffraction pattern from samples annealed at 300 °C, 1h. (b) AES depth profiles from samples in (a). (c) TEM micrograph and diffraction pattern from samples heat-treated at 300 °C for 45 h.

59

the diffraction pattern mode. Crystalline β-Co was formed , as shown in Fig. 1(a), after annealing at 300 °C for 1 h. The β-Co film is highly oriented with respect to the GaAs substrate. No compound phase was detected by TEM. However, AES depth profiles, as seen in Fig. 1(b), revealed that a significant amount of outward diffusion of Ga occurred. After subsequent annealing at 300 °C for 15-45 h. , the Co films were found to react with GaAs to form $Co(Ga_{1-x}As_x)$ which is considered to be a solid-solution of CoAs. A typical diffraction pattern and a dark field image are shown in Fig. 1(c). $Co(Ga_{1-x}As_x)$ grains oriented with respect to (001) GaAs are clearly evident. Genut and Eizenberg[23] and Palmstrom et al.[24] also reported the formation of Co_2GaAs which they considered to be a ternary phase.

A complicated TEM diffraction pattern exhibiting the coexistence of textured, untextured and amorphous phases was observed from samples annealed at 400 °C as seen in Fig. 2. The phase $Co(Ga_{1-x}As_x)$ started to decompose into CoGa and CoAs, and Ga diffused out to react with oxygen to form a $β-Ga_2O_3$ surface layer. The oxide was amorphous at 400 °C but became polycrystalline when annealed at higher temperatures.

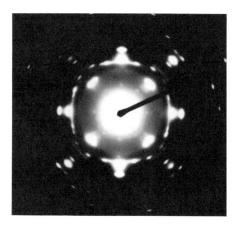

Fig. 2. TEM diffraction pattern from samples annealed at 400 °C, 1 h.

The AES depth profiles (displayed in Fig. 3) indicate that the interfacial reactions are dominated by the inward diffusion of cobalt. This is evident from the carbon atoms, originally located at Co/GaAs interface, which have been dispersed away from the Co/GaAs interface in the samples annealed at 250 - 300°C. These carbon atoms moved toward the oxide/reacted layer interface in the samples annealed at 400 -700 °C. Cobalt diffuses through the carbon layer to react with gallium arsenide. Gallium must also diffuse outwards through the reacted layer to the surface to form $β-Ga_2O_3$. However, arsenic was found to be essentially immobile and stay within the reacted layer at temperatures below 700 °C. Comparing the Auger peak intensity ratio of Ga to As from the substrate with those from the reacted layer as shown in Fig.3(c) , it is evident that the concentration of As is much higher than that of Ga in the reacted region apparently as a result of Ga out-diffusion.

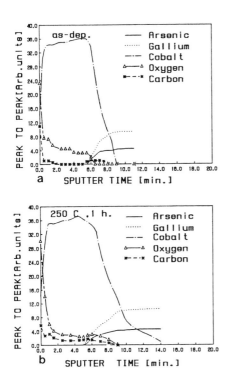
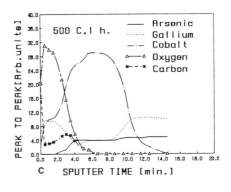

Fig. 3. AES depth profiles obtained from samples. (a) un-annealed, and annealed at (b) 250 °C, 1 h. (c) 500 °C, 1 h.

ESCA equipped with Ar+ sputtering was used to study the reacted layer in the samples annealed at 600 °C. The chemical shifts of Ga 3d and Co $2P_{3/2}$ in the ESCA binding energy spectra are shown in Fig.4. The shift of binding energy for the Ga 3d peak from the surface to the substrate was determined to be 20.7 eV --> 18.6 eV --> 19.1 eV which corresponds to β-Ga_2O_3 --> CoGa --> GaAs. The binding energy values for Ga 3d from β-Ga_2O_3 and GaAs were reported in the literatures to be 20.6 eV [25] and 19.1 eV [26], respectively. However, no appreciable shift of Co 2P $_{3/2}$ peak was detected. On the basis of TEM diffraction pattern analysis with Auger depth profiles and ESCA data, it is considered that a layered structure of "β-Ga_2O_3/ (CoGa,CoAs) /GaAs" was formed.

In the 500 °C annealed samples, both CoGa and CoAs were found to be preferentially oriented with respect to the (001) GaAs substrate as evident from the diffraction pattern shown in Fig.5(a). The orientation relationships were determined to be :

[001] CoGa // [001] GaAs, (220) CoGa // (220) GaAs
[101] CoAs // [001] GaAs, (020) CoAs // (220) GaAs

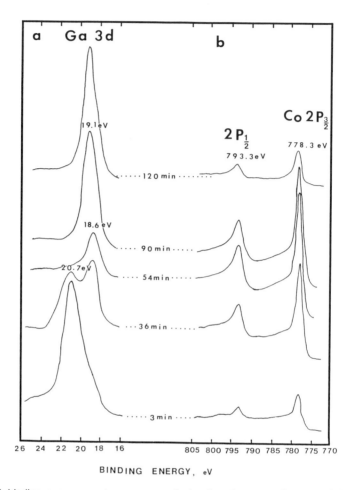

a   Ga 3d   b

Co 2P₃/₂

2 P₁/₂
793.3eV   778.3 eV

19.1 eV

····120 min·····

18.6 eV

····90 min····

20.7 eV

····54min····

···36min······

····3min····

26 24 22 20 18 16   805 800 795 790 785 780 775 770

BINDING ENERGY, eV

Fig. 4 ESCA binding energy spectra versus sputtering time from samples annealed at 600 °C , 1 h. (a) Ga 3d , β–Ga₂O₃ (20.7 eV) ----> CoGa (18.6 eV)----> GaAs (19.1 eV) (b) Co 2P₃/₂ , no appreacible peak shift detected.

Fig. 5(b) shows the Moire fringes from regions as large as 2 µm in size for a specimen taken with (020) CoAs (d₁ = 0.1729 nm) and (220) GaAs (d₂ = 0.1998 nm) diffraction spots.   The average spacing was measured to be 1.25 nm which is very close to the calculated value of (D) 1.28 nm where (D) = d₁d₂/d₁-d₂ . The layered structure is also evident in the cross-sectional TEM micrograph of this sample shown in Fig. 5(c).

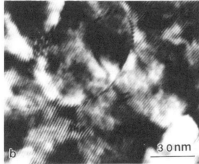

Fig. 5. (a) TEM diffraction pattern from samples annealed at 500 °C for 1 h. (b) The Moire fringes obtained by taking (020) CoAs and (220) GaAs diffraction spots simultaneously from samples (a). (c) Cross-sectional image showing the layered structure from samples (a).

CoGa was found to grow epitaxially on (001) GaAs. In samples annealed at 600 - 700 °C, epitaxial regions of 1-2 μm were found. Selected area diffraction (SAD) combined with EDS analysis confirmed the presence of CoGa epitaxy. Regular interfacial dislocations were observed. The average dislocation spacing was measured to be 15 nm. The best epitaxy was obtained in samples heat-treated first at 250 °C for 1 h. followed by a subsequent annealing at 750 °C for 1 h. Epitaxial regions as large as 6 μm were observed. An example is shown in Fig. 6. Almost all of the CoGa regions were found to be epitaxially related to the substrate.

Fig. 6. Composite TEM micrograph obtained in samples heat treated by a 250 - 750 °C two-step annealing technique showing elongated CoGa epitaxy up to 6 μm in length.

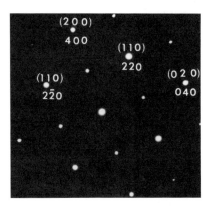

Fig. 7. The overlapping [001] CoGa/[001] GaAs TEM diffraction pattern from the CoGa regions in Fig.6; CoGa and GaAs diffraction spots are presented by indices with and without brackets, respectively.

Fig. 7 shows the overlapping [001] CoGa/[001] GaAs diffraction pattern. The orientation relationships were determined to be [001] CoGa // [001] GaAs and (220) CoGa // (220) GaAs. The interfacial dislocation networks are best imaged by the weak beam dark field image technique. The image from the [220] GaAs diffraction spot is shown in Fig. 8(a). The interfacial dislocation networks can be resolved into two sets of parallel dislocation fringes perpendicular to each other by [040] and [400] GaAs weak beam dark field images, as shown in Figs. 8(b) and (c), respectively. Moire fringes are also present in the dark field images taken with [220], [040] and [400] GaAs diffraction spots, as seen in Figs. 8(d), (e) and (f), respectively.

XTEM micrographs revealed that the interfacial dislocation networks are three-dimensional in character, as seen in Fig. 9(a). The Burgers vectors of interfacial dislocations were identified to be 1/2 <101> and 1/2 <011> which are inclined to the (001) GaAs surface. The segments of dislocations are along [001], <101> and <011> GaAs directions with the projected dislocation images viewed along [001] GaAs direction being along [100] and [010] GaAs directions.

No interfacial dislocations were observed in most of epitaxial CoAs, 1-2 µm in size,which is considered to be pseudomorphic with GaAs as shown in Fig. 10(a). The overlapping [010] CoAs // [001] GaAs diffraction pattern and the indexed pattern are shown in Fig. 10(b) and (c), respectively. The orientation relationships between CoAs and GaAs were determined to be [101] CoAs // [001] GaAs and (020) CoAs // (220) GaAs. It is to be noted that interfacial dislocations were occasionally observed in some of the CoAs regions as shown in Fig. 10(d).

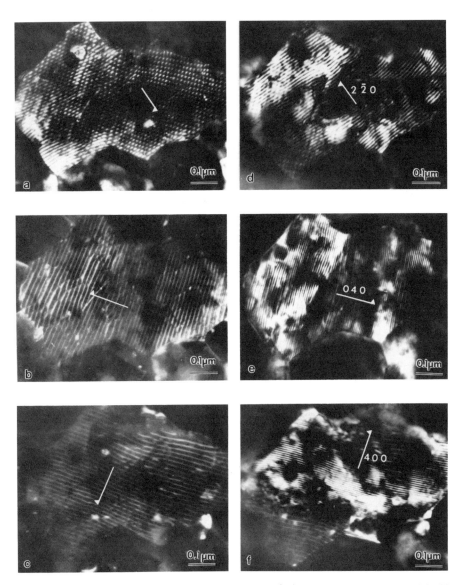

Fig. 8. TEM micrographs from the same region in 750 °C, 1 h. annealed samples. (a), (b) and (c) are weak beam dark field images taken with the [220], [040] and [400] GaAs diffraction spots, respectively, which show that the dislocation networks in (a) are resolved into two set of parallel dislocation fringes (b) and (c); (c), (d) and (e) are dark field images taken with [220], [040] and [400] GaAs diffraction spots respectively, showing the presence of CoGa Moire fringes.

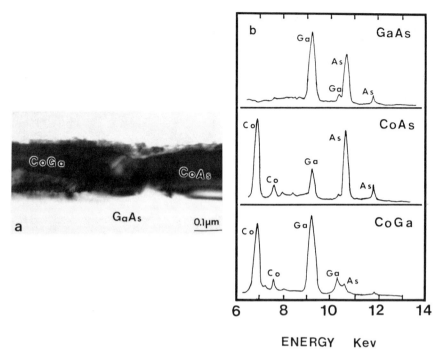

ENERGY   Kev

Fig. 9. (a) Cross-sectional TEM micrograph of samples annealed at 800 °C for 1 h. showing that the interfacial dislocation networks present at CoGa/GaAs interface are three-dimensional in character. (b) EDS spectra taken from the marked regions GaAs, CoAs and CoAs in (a). The Ga peak is apparently present in the CoAs region. No As signl is observed in the CoGa region.

EDS analysis of selected regions of the cross-sectional samples annealed at 800 °C show that appreciable amounts of Ga are present in regions which exhibit diffraction patterns corresponding to those of CoAs, as seen in Fig. 9(b). The regions are therefore considered to the solid solution of CoAs and Ga , $Co(Ga_{1-x}As_x)$.

Discussion

As pointed out previously by Lin et al.[16], phases with compositions lying on GaAs-M connection line are favored to form first when metal/GaAs contacts are brought up to a sufficiently high temperature. This has been demonstrated experimentally for Ni/GaAs[20,22] and Pd/GaAs [15]. In the case of Ni/GaAs, a ternary phase is formed initially while a binary solid solution $Pd_5(Ga_{1-x}As_x)_3$ with extensive solubility for Ga is formed in the case of Pd/GaAs. In the present case, it is believed that the phase formed initially is a solid solution of CoAs as represented by $Co(Ga_{1-x}As_x)$, analogous to the case of Pd/GaAs. EDS analysis of a sample annealed at 800°C for 1 h. as exhibited in Fig. 9(b) shows that this phase exists at this temperature. However, further phase diagram and bulk-diffusion couple studies are needed to substantiate this finding.

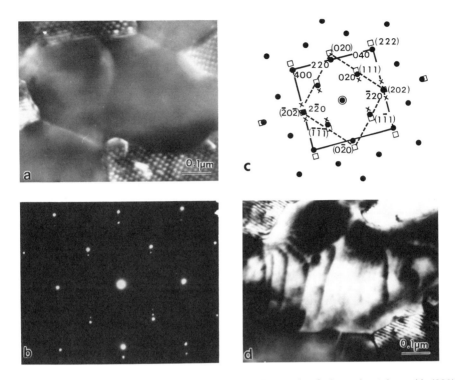

Fig.10. (a) Weak beam dark field image corresponding to the CoAs region taken with (220) GaAs diffraction. (b) Diffraction pattern from (a). (c) The indexed pattern, (□) CoAs, (●) GaAs, (x) double diffraction pattern. (d) Interfacial dislocations observed in some of the CoAs regions.

Arsenic sublimation has been a serious problem in M/GaAs samples annealed at high temperature. Palmstrom [24] reported that, due to substantial As loss from the reacted layer, epitaxial CoGa was the only phase observed in samples annealed at 750 °C for 1 h. No serious loss of As appeared in our samples even when annealed at 800 °C for 1 h. as inferred from the AES depth profiles and EDS analysis of cross-sectional TEM specimens, given in Figs. 3 and 9. This may be attributed to our use of encapsulated samples as well as the formation of a β-Ga_2O_3 surface layer which minimized the As loss from surface.

The tendency of the outward diffusion of Ga to form β-Ga_2O_3 on the surface is in part related to the higher thermodynamic stability of this compound. It is more stable than any cobalt or arsenic oxides. Similar results were previously observed in the Pt/GaAs[27], Au/GaAs[28] and Ni/GaAs[29] systems.

Zur et al.[30,31] have pointed out that the smaller the common-unit cell dimension and the less the mismatch, the more likely is the epitaxial growth of the overlayer on the substrate. Lattice matches in epitaxial CoGa and CoAs on (001) GaAs systems are listed in Table I. The fact that CoGa has a similar crystal structure, small common-unit cell dimension and relatively small mismatch with respect to (001) GaAs may explain why almost all of CoGa regions are epitaxially related to the substrate. Once epitaxial CoGa

Table I. Lattice matches in CoGa and CoAs on (001) GaAs systems; a and b are the sides of the common-unit cell; α is the acute angle between them.

Phases	Matching Plane	Epitaxial Condition	Cell-Area (A^2)	CoGa or CoAs a	b	α	GaAs a	b	α	Mismatches;% a	b	α
CoGa	(001)/(001)GaAs	[220]/[220]	16	4.07	4.07	90	4.0	4.0	90	1.85	1.85	0
CoAs	(101)/(001)GaAs	[020]/[220]	234	11.26	20.75	90	11.99	19.99	90	-6.28	3.73	0

has formed, the CoAs grows epitaxially, but forms smaller regions. A two-step annealing was found to be effective in promoting the epitaxial growth of CoGa and CoAs on GaAs as is the case for many silicide/silicon systems[32]. It is believed that the first annealing step is to drive impurities away from the Co/GaAs interface. Thus, fewer nucleation sites at the interface favor epitaxial growth. AES depth profiles as displayed in Fig. 3(b) indeed show that carbon atoms, originally located at the Co/GaAs interface, have been displaced outwards in samples after a 250 ^oC, 1 h. anneal.

Summary and Conclusions

The initial phase formed at low temperature was $Co(Ga_{1-x}As_x)$ which is considered to be the solid solution of CoAs and Ga. This phase subsequently decomposed into CoGa and CoAs upon subsequent annealing. The completely reacted layer from a Co/GaAs couple is "β-Ga_2O_3/ (CoGa, CoAs) /GaAs". The formation of a β-Ga_2O_3 surface layer and the use of encapsulated samples are believed to minimized As loss from the reacted layer.

Both CoGa and CoAs are found to grow epitaxially on (001) GaAs. The orientation relationships between CoGa and GaAs were determined to be [001] CoGa // [001] GaAs and (220) CoGa // and (220) GaAs. The Burgers vectors of interfacial dislocations were identified as 1/2 <101> and 1/2 <011> which are inclined to (001) GaAs surface. No interfacial dislocations were observed in most of epitaxial CoAs, which is considered to be pseudomorphic to GaAs. The orientation relationships between CoAs and GaAs were determined to be [101] CoAs // [001] GaAs and (020) CoAs // (220) GaAs. Two-step annealing was effective in promoting epitaxial growth. CoGa was found to readily grow epitaxially on (001) GaAs. It is, therefore, a most promising candidate phase for forming single crystalline contacts to GaAs by codeposition.

Acknowledgements

The authors would like to thank the Department of Energy for financial support through Grant DE-FG02-86ER452754 and Dr. Joe Darby, Jr. of DOE for his interest in this work. Valuable discussions with Dr. S. E. Babcock , Dr. J. -C Lin , K. J. Schulz, K. C. Vlack, R. A. Konetzki , and D. A. Sluzewski are also gratefully acknowledged.

Reference

1. A. K. Sinha and J. M. Poate, in Thin Films - Interdiffusion and Reactions, edited by J. M. Poate, K. N. Tu, and J. W. Mayer (Wiley, New York, 1978), p407 and references therein.
2. C. J. Palmstrom and D. V. Morgan, in Gallium Arsenide : Materials, Devices, and Circuits, edited by M. J. Howes and D. V. Morgan (Wiley, New York, 1985), p.195.
3. T. S. Kuan, P. E. Batson, T. N. Jackson, H. Rupprecht, and E. L. Wilkie, J. Appl. Phys. 54, 6952 (1983).
4. W. T. Anderson, Jr., A. Christon, and J. E. Davey, J. Appl. Phys. 49, 2998 (1978).
5. G. Y. Robinson, Solid-state Electron. 18, 331 (1975).
6. A. K. Sinha and J. M. Poate, Appl. Phys. Lett. 23, 666 (1973).
7. V. Kumar, J. Phys. Chem. Solid 36, 535 (1975); C. Fontaine, T. Okumara, and K. N. Tu, J. Appl. Phys. 54, 1404 (1983).
8. K. M. Yu, J. M. Jakievic, and E. E. Haller, Mater. Res. Soc. Symp. Proc. 69, 281(1986).
9. T. Sands, V. G. Keramidas, A. J. Yu, K. M. Yu, R. Gronsky, and J. Washburn, J. Mater. Res. 2 , 262 (1987).
10. J. O. Olowolafe, P. S. Ho, H. J. Hovel, J. E. Lewis, and J. M. Woodall, J. Appl. Phys. 50, 955 (1979).
11. X.-F. Zeng and D. D. L. Chung, J. Vac. Sci. Technol. 21, 611 (1982).
12. A. Oustry, M. Caumont, A. Escaut, A. Martinez, and B. Toprasertpong, Thin Solid Films 79 , 251 (1981).
13. T. S. Kuan, J. L. Freeouf, P. E. Baston, and E. L. Wilkie, J. Appl. Phys. 58, 1519 (1985).
14. T. Sands, V G. Keramidas, R. Gronsky, and J. Washburn, Mater. Lett 3, 409 (1985); T. Sands, V. G. Keramidas, A. J. Yu, K. M. Yu, R. Gronsky, and J. Washburn, Mater. Res. Soc. Symp. Proc. 54, 367 (1986).
15. J.-C. Lin, K.-C. Hsieh, K. J. Schulz and Y. A. Chang , J. Mater. Res.,1988,in press.
16. J.-C. Lin, K.J. Schulz , K.-C. Hsieh and Y.A. Chang, J. Electrochem. Soc.,1987, under review.
17. M. Ogawa, Thin Solid Films 70, 181 (1980).
18. A. Lahav, M. Eizenberg, and Y. Komen, Mater. Res. Soc. Sym. Proc. 37, 641 (1985).
19. T. Sands, V. G. Keramidas, J. Washburn, and R. Gronsky, Appl. Phys. Lett. 48, 402 (1986).
20. J.-C. Lin, X.-Y. Zheng, K.-C. Hsieh and Y.A. Chang, in MRS Proceedings on Epitaxy of Semiconductor Layered Structures. (Eds. T. Tung , L.r. Dawson and R. L. Gunshov), Mater. Res. Soc., Pittsburgh. PA15237,1988,1n press.
21. A. J. Yu, G. J. Galvin, C. J. Palmstrom, and J. W. Mayer, Appl. Phys. Lett. 47, 934 (1985).
22. T. Sands, V. G. Keramidas, K. M. Yu, J. Washburn, and K. Krishnan, J. Appl. Phys. 62, 2070 (1987).
23. M. Genut and M. Eizenberg. Appl. Phys. Lett. 50, 1358 (187).
24. C. J. Palmstrom, C. C. Chang, A. J. Yu, G. J. Galvin, and J. W. Mayer, J. Appl. Phys. 62 , 3755 (1987).
25. G. Schon , J. Elect. Spectros.2,75(1975).
26. T. Lave, C. J. Vesely, and D. W. Langer, Phys. Rev. B6,3770(1972).
27. C.C. Chang , S.P. Murarka, V. Kumar ,and G. Quintana, J. Appl. Phys. 46,4237 (1975).
28. G.Y. Robinson , J. Vac. Sci. Technol., 13,884 (1976).
29. L.J. Chen and Y.F. Hsieh. Mater. Res. Soc. Symp. Proc. 31,165 (1984).
30. A. Zur and T. C. McGill, J. Appl. Phys. 55, 378 (1984).
31. A. Zur, T. C. McGill, and M. A. Nicolet, J. Appl. Phys. 57, 600 (1985).
32. L. J. Chen, H. C. Cheng, and W.T. Lin, in Materials Research Society Symposium Proceedings 54, edited by R. J. Nemanich, P. S. Ho, and S. S. Lau (Materials Reasearch Society , Pittsburgh,1986) p.245.

Session II:

MICROSTRUCTURE OF THIN FILMS

Session Chairman
David A. Smith
IBM T. J. Watson Research Center
Yorktown Heights, New York

THEORY OF ORIENTATION TEXTURES DUE TO

SURFACE ENERGY ANISOTROPIES

Jean E. Taylor and John W. Cahn

Department of Mathematics
Rutgers University, New Brunswick, NJ 08903

Institute for Materials Science and Engineering
National Bureau of Standards, Gaithersburg, MD 20899

Abstract

The effect of surface free energy on the orientation of nuclei on a substrate is examined. In general, the orientation of the nuclei is not determined by minimizing the nucleus-substrate interfacial free energy alone: all interfacial anisotropies must be considered. That orientation preferences should result from anisotropies in the substrate-nucleus interfacial free energies is obvious, but we find that strong orientation effects are present even if this energy is assumed to be independent of the nucleus orientation, provided that the nutrient-nucleus interfacial free energies are anisotropic. Furthermore, varying the magnitude of even an isotropic nucleus-substrate energy will change the orientation of the nuclei. The theory points to some simple rules for prediction of nucleation orientations.

Microstructural Science for Thin Film
Metallizations in Electronics Applications
Edited by J. Sanchez, D.A. Smith and N. DeLanerolle
The Minerals, Metals & Materials Society, 1988

We address the question: how does surface free energy anisotropy cause certain orientations to dominate during the formation of thin films? We report on the nucleation stage, showing how the barrier height for nucleation on a substrate depends on orientation and give an explicit formula for the orientation which has the least barrier height. The proofs of these results, as well as the consideration of other aspects of this problem (such as how competition during growth and coarsening selects grains of particular orientations), we will give in a later paper.

The crystal-substrate interfacial energy is an obvious possible anisotropic factor, since it can depend on the orientation of the crystal. It enters the theory as a difference, that we will occasionally refer to as the wetting strength. Let $\Delta\sigma = \sigma_{CS} - \sigma_{SM}$, where CS stands for crystal-substrate, SM stands for substrate-medium, and both σ's are free energies per unit area; we must subtract σ_{SM} since part of the medium-substrate interface is being replaced by the crystal-substrate interface. Note that $\Delta\sigma$ can be negative.

Some general conclusions to be drawn from the nucleation formula given below are that one must consider anisotropies of the crystal-medium surface free energy σ as well as those of $\Delta\sigma$, and that one must also consider their relative magnitudes. An additional result is that as the magnitude of the wetting strength is varied without changing its orientation dependence, abrupt changes in orientation are predicted and that for some range of this parameter asymmetric orientations can have the lowest barrier heights. In one of the two-dimensional cases mentioned below, every orientation has the lowest nucleation barrier for some value of the wetting strength.(1)

We note that often σ and $\Delta\sigma$ are not known; however, experimental evidence can be coupled with this theory to measure approximately these quantities. To the extent that parameters such as wetting strength can be varied, the orientation of nuclei could then be controlled.

Wulff's construction gives the equilibrium crystal shape W_σ for a crystal in a medium, at the scale such that the total surface free energy of the Wulff shape is n times its volume, where n is the ambient dimension (in this paper usually n is 2 or 3). In terms of a mathematical formula (2),

$$W_\sigma = \{\mathbf{x} : \mathbf{x} \cdot \nu \leq \sigma(\nu) \quad \text{for every unit vector } \nu\}.$$

This construction was modified by Winterbottom (3) to give the shape of a crystal of a given orientation on a substrate. The Winterbottom shape is the Wulff shape chopped off by one more half space, that of the substrate translated by the signed distance $\Delta\sigma$ from the origin. See figure 1. If $\Delta\sigma$ is sufficiently large and positive, then nothing is chopped off and the crystal will not form on the substrate; if $\Delta\sigma$ is sufficiently large and negative, then the crystal will completely wet the interface, nucleating without a barrier and (at equilibrium on an infinite substrate) spreading out in a thin film.

To consider several different possible orientations, it is much more convenient for us to leave the orientation of the crystal fixed and change the orientation of the substrate; $\Delta\sigma$ becomes a function of the orientation of the substrate. Since we are looking only for the least energy configurations, we can let $\Delta\sigma$ be a function of unit vectors only, defining $\Delta\sigma(\mathbf{n})$ to be the smallest value of $\Delta\sigma$ among all rotations about the axis $\mathbf{n}$. (Note that each of σ_{CS} and σ_{SM} might depend on $\mathbf{n}$.)

When the nutrient medium is metastable with respect to the formation of a crystal, there is a competition between the energy decrease due to the new phase (proportional to the volume of new phase created) and the energy increase due to surface energy. A crystal of a given volume and given orientation will have the lowest total energy when its

Figure 1 - The Winterbottom modification of the Wulff construction for the case where the Wulff shape is a square centered at the origin, the orientation of the substrate is **n**, and the value of $\Delta\sigma(\mathbf{n})$ is s (if s is negative, the crystal on the substrate is the shaded part; if s is positive, the crystal on the substrate is all of the square except for the shaded part). The circle shows that the midpoint of the slice is $s\mathbf{n}$ itself. Thus in case $\Delta\sigma$ has the value s for all orientations of the substrate, the area of the crystal on the substrate is a local maximum or minimum.

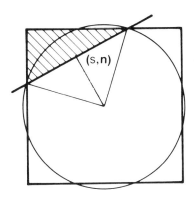

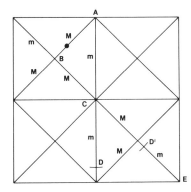

Figure 2 - The locus of all points $s\mathbf{n}$ such that the area cut off by a substrate of orientation **n** at signed distance s from the origin is a local maximum or minimum, in the case where $\Delta\sigma$ is constantly s (and the Wulff shape is again a square centered at the origin). The parts labelled M give local maxima when s is positive and local minima when s is negative; the parts labelled m do the opposite. The points labelled B,C,D,and D' are where the orientation changes discontinuously with the wetting strength parameter s. The point labelled A is the limiting orientation as the wetting strength reaches the limit of complete wetting, and the point E is the limiting orientation as the orientation reaches the limit of complete drying. The fat dot is the point $s\mathbf{n}$ corresponding to the Winterbottom shape of figure 1.

Figure 3 - The locus of all points $s\mathbf{n}$ such that the area cut off by a substrate of orientation **n** at signed distance s from the origin is a local maximum or minimum, in the case where $\Delta\sigma$ is $s\sigma$. The labelling is as in figure 2.

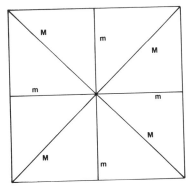

shape is that of the Winterbottom shape, since that shape has the least surface energy for the volume it contains.

Furthermore, if the size of the nucleus crystal has a scale factor λ compared to the Winterbottom construction, then the total free energy change due to its formation is

$$\Delta F = -\omega \lambda^n volume(W) + n\lambda^{n-1} volume(W),$$

where n is again the dimension of space, ω is the free energy decrease per unit volume, and W is the Winterbottom shape. (Recall that the surface energy of the Winterbottom shape is exactly n times its volume.(4)) The barrier height is the maximum value of ΔF and occurs (using calculus) at $\lambda = (n-1)/\omega$; the maximum value is $((n-1)/\omega)^{n-1} volume(W)$. Thus the height of the energy barrier is proportional to the volume of W, and the orientation of the crystal on the substrate which has the lowest barrier to nucleation is the one with the smallest volume for its Winterbottom shape.

The problem of determining the lowest barrier orientation thus reduces to finding the orientation $\mathbf{n}$ of the substrate for which the Winterbottom construction has the smallest possible volume. Even when $\Delta\sigma$ is constant, this orientation is usually unique (up to the symmetry of the underlying Wulff shape in the absence of any substrate); furthermore, the orientation depends on the *value* of that constant.

One two-dimensional example, demonstrating some interesting consequences, has been considered in detail: when the Wulff shape is a square in the plane and $\Delta\sigma$ is assumed isotropic, all anisotropy comes from σ alone, even though all crystal-nutrient surfaces have the same energy. The orientation with the lowest nucleation barrier was found to vary with the assumedly isotropic wetting strength; EVERY orientation gives a minimum for some value of the wetting strength. The optimum orientation as a function of wetting exhibited two kinds of bifurcations (abrupt discontinuities in the functional behavior of the orientation.) This example is discussed in detail in (1) and is illustrated in Figure 2.

The general result is as follows:

An orientation $\mathbf{n}$ is *critical* (giving a minimum, maximum, or saddle point) if and only if the *centroid* of the slice of the Wulff shape by the plane $\mathbf{n} \cdot \mathbf{x} = \Delta\sigma$ is the point $\xi_{\Delta\sigma}(\mathbf{n})$, where

$$\xi_{\Delta\sigma}(\mathbf{n}) = grad(|\mathbf{m}|\Delta\sigma(\mathbf{m}/|\mathbf{m}|)), \text{evaluated at the point } \mathbf{m} = \mathbf{n}.$$

(The ξ notation is used in analogy with the ξ vector of Cahn and Hoffman (5,6), which is the gradient of σ (similarly extended to be a function on all vectors rather than just unit vectors).) Note that when $\Delta\sigma$ is constant, then $\xi_{\Delta\sigma}(\mathbf{n})$ is just $\mathbf{n}$ times that constant.

As another example of this formula, consider the very special case where the Wulff shape is a polyhedron, σ has the smallest possible values it can have and still produce the same Wulff shape (so that the polar plot of σ is composed of spheres through the origin), and $\Delta\sigma$ is a constant s times σ. Then wherever $\xi_{\Delta\sigma}$ is defined, its value is a corner of the Wulff shape. It is not defined on the unit normal vectors of the faces of the Wulff shape, nor on the great circle segments corresponding to edges of the Wulff shape which connect these unit vectors. On the latter, however, the "gradient" in some generalized sense is constrained to lie on the line between the corners of the Wulff shape which are the endpoints of the edge (by the assumption that $\Delta\sigma$ is just s times σ). Thus one can show that the "gradient equals the centroid" condition here selects out one vector for each edge of the Wulff shape, and that vector is independent of the value of s. (For the Wulff shape being a cube, that vector is a (110) direction.) Thus in this special case, it is not true that every orientation gives a minimum for some value of s. Furthermore, the orientations which give a minimum in the limiting cases of complete wetting or complete "drying" can be quite different from those in the case where $\Delta\sigma$ is a constant. See Figure 3.

We reiterate: the general formula stated above gives the orientations of crystals on a substrate which have the lowest barriers for nucleation. Examples show that there are sizable differences in barrier heights even when $\Delta\sigma$ is isotropic (i.e. a constant), provided that σ is anisotropic, and that the orientation which has the lowest barrier height can vary dramatically with the scale factor of $\Delta\sigma$. Asymmetric orientations can be minimizing for many functions $\Delta\sigma$, so the presence of asymmetric orientations in experiments need not be regarded as a fluke. Further experimental evidence of preferred orientations, especially under conditions where $\Delta\sigma$ can be varied (e.g. by substances adsorbed on the substrate), would be useful.

We would like to acknowledge the partial support of NSF and AFOSR (for the first author) and ONR (for the second author).

REFERENCES

1. J. W. Cahn and J. E. Taylor, "The Influence of Equilibrium Shape on Heterogeneous Nucleation Textures," Phase Transformations '87, The Institute of Metals, to appear.

2. J. E. Taylor, "Crystalline Variational Problems," Bull. Amer. Math. Soc. **84** (1978) 568-588.

3. W. L. Winterbottom, "Equilibrium Shape of a Small Particle in Contact with a Foreign Substrate," Acta Met. **15** (1967) 303-310.

4. J. K. Lee and H. I. Aaronson, "Application of the Modified Gibbs-Wulff Construction to Some Problems in the Equilibrium Shape of Crystals at Grain Boundaries," Scripta Met. **8** (1974) 1451-1460.

5. D. W. Hoffman and J. W. Cahn, "A Vector Thermodynamics for Anisotropic Surfaces – I. Fundamentals and Applications to Plane Surface Junctions," Surf. Sci. **31** (1972) 368-388.

6. J. W. Cahn and D. W. Hoffman, "A Vector Thermodynamics for Anisotropic Surfaces – II. Curved and Facetted Surfaces," Acta Met. **22** (1974) 1205-1214.

COMPUTER SIMULATION OF MICROSTRUCTURAL EVOLUTION IN THIN FILMS

H. J. Frost* and C. V. Thompson**

*Thayer School of Engineering, Dartmouth College, Hanover, N.H. 03755
**Dept. of Materials Science and Engineering, M.I.T., Cambridge, MA 02139

Abstract

The nature of the microstructure of a thin film strongly affects its functionality in electronic applications. For example, the rate of electromigration-induced failure is a function not only of the grain size in an interconnect line, but also of the width and shape of the grain size distribution. We are developing techniques which allow prediction of the relationships between the conditions for thin film processing and the topology and geometry of resulting grain structures. We have simulated two-dimensional microstructural evolution by determining the location of grain boundaries after nucleation and growth of crystalline domains. We have allowed for nucleation under a variety of conditions including constant nucleation rates, decreasing nucleation rates and instantaneous saturation of nucleation sites. We have also allowed for increasing and decreasing growth rates which depend in various ways on the domain size. We have simulated grain growth in two-dimensional structures by allowing boundary and triple point motion in order to reduce the total grain boundary area. The rate and nature of the initial stages of grain growth are strongly affected by the conditions for nucleation and growth. Eventually, however, grain growth appears to proceed as expected from analytical treatments.

Microstructural Science for Thin Film
Metallizations in Electronics Applications
Edited by J. Sanchez, D.A. Smith and N. DeLanerolle
The Minerals, Metals & Materials Society, 1988

Introduction

Polycrystalline films are used in a wide variety of applications, including as conductors [1], and occasionally as active materials [2], in electronic devices and circuits, as magnetic storage media [3], as mechanical elements in integrated sensors [4], and as protective coatings. [5] In all of these applications, the properties of polycrystalline films are strongly affected by their microstructures, i.e., the average grain size, grain orientations and the grain size distributions. For example, the dominant failure mechanism for thin film conductors, contacts and interconnects, is electromigration. Electromigration occurs along grain boundaries and failure initiates at grain boundary surface intercepts. [6] It has been proposed that the distribution of electromigration-induced times to failure are directly related to grain size distributions. [7] It has been clearly demonstrated that mean times to failure are a function of grain size [5,8] and geometry [9].

Motivated in part by recognition of the wide spread importance of microstructure-sensitive applications, there have been an increasing number of experimental and theoretical studies of microstructural evolution in thin films, as indicated, for example, by the papers presented in this symposium. The initial microstructure develops during the early stages of deposition of a crystalline film or during crystallization of an amorphous film. We have modelled the evolution of two-dimensional microstructures under various conditions of nucleation and growth [10, 11] and will review these results below. In general we have found that the geometry of the initial microstructures depends strongly on the conditions for nucleation and growth. The geometric differences persist even during the early stages of normal grain growth [12], as will also be discussed below, though steady-state, initial-structure-independent grain growth eventually occurs.

On the basis of experiments it is known that thin film normal grain growth generally does not result in grain sizes significantly larger than the film thickness, and it is still likely to be sensitive to the geometry of the initial as-deposited or as-crystallized microstructure. [13] Also, in thin films, abnormal or secondary grain growth can play an important or dominant role in determining the final grain sizes and orientations. [13] In this paper we review our recent results for computer modelling of these phenomena which are known to affect microstructures of thin films; nucleation and growth to form initial microstructures, the early stages of normal grain growth, and the early stages of secondary grain growth.

Nucleation and Growth to Impingement Microstructures

The microstructures that result from nucleation and growth to impingement depend on the distribution of nucleation sites in both space and time, and on the growth velocity of the grains after nucleation. There are an infinite number of different conditions of nucleation distribution and growth rates possible, which would produce different microstructures. We have concentrated on a relative few simple cases. For one set of cases we have assumed that the growth rate is constant with time for each grain, and identical for all grains. With this constant growth rate we have used various nucleation conditions to produce different microstructures. For another set of conditions we have held the nucleation conditions constant, but have taken the growth rate of individual crystals to be a function of the crystal radius. For different growth-rate functions, we obtain different microstructures.

Effect of Nucleation Conditions

The simplest nucleation condition is the simultaneous nucleation at a fixed number of sites, randomly distributed on the plane, which we call the site saturation case. This case applies when a fixed number of heterogeneous nucleation sites are quickly saturated before any significant crystal growth occurs. It produces the well-studied Voronoi polygon structure. (Fig. 1c) The Voronoi polygon for a particular nucleation point is that region of the plane that is closer to that point than to any other nucleation point. It is also known as a Wigner-Seitz cell, a Dirichlet region, or a Theissen polygon. Studies of this structure have been reviewed by Getis and Boots. [14]

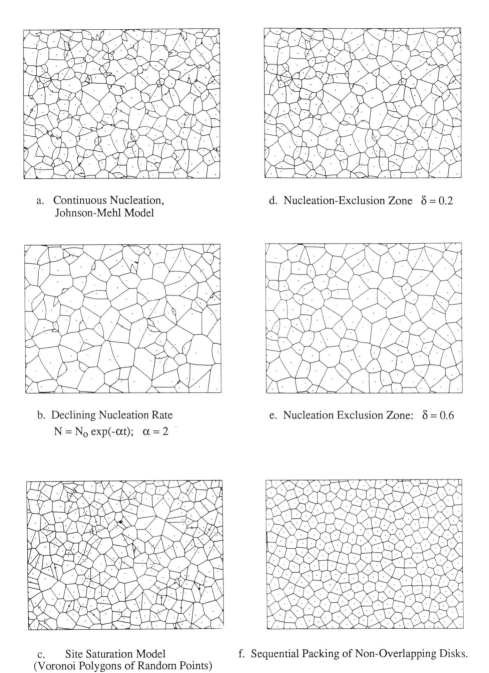

a. Continuous Nucleation,
   Johnson-Mehl Model

d. Nucleation-Exclusion Zone $\delta = 0.2$

b. Declining Nucleation Rate
   $N = N_0 \exp(-\alpha t);\quad \alpha = 2$

e. Nucleation Exclusion Zone: $\delta = 0.6$

c.   Site Saturation Model
   (Voronoi Polygons of Random Points)

f. Sequential Packing of Non-Overlapping Disks.

Figure 1 - Structures resulting from different nucleation conditions, with a constant growth rate.

Another simple nucleation condition is continuous nucleation, in which the nucleation rate, per unit of untrans-formed area, remains constant throughout the process of film formation. This case applies for two-dimensional homogeneous nucleation . If the nucleation sites are randomly distributed in the plane, and the growth rate is constant and isotropic for all crystals, the resulting structure is that first described by Johnson and Mehl [15] and Evans [16]. (Fig. 1a) It has been less thoroughly studied than the Voronoi polygon structure, but various properties have been calculated by Meijering [17] and by Gilbert [18]. The most notable difference between simultaneous nucleation and continuous or sequential nucleation is that the former leads to straight boundaries and for the later the boundaries are segments of hyperbolae. Structures similar to the continuous nucleation structure (including hyperbolic grain boundaries) have been observed in experiments involving the crystallization of thin films of amorphous $CoSi_2$. [19,20]

Intermediate between continuous nucleation and site saturation are a series of cases in which the nucleation rate decreases with time, modelling the progressive exhaustion of a finite number of sites. (Fig. 1b) We assume that the probability of nucleation at a given site, per unit time, remains constant. The nucleation rate varies with time according to the number of sites remaining where nucleation has not yet occurred. If the density of remaining sites is N, the nucleation rate is given by $-dN/dt = \alpha N$, where α is a constant. The density or remaining sites follows an exponential decay with time: $N(t) = N_0 \exp(-\alpha t)$, where N_0 is the initial site density. The nucleation rate also follows an exponential decay with time. In the limit $\alpha \rightarrow 0$, we have continuous nucleation. In the limit that α is infinite, we have site saturation. Figure 2 shows the variation with time of the nucleation rate (per unit of untransformed area) for various values of α.

In an alternative set of conditions, nucleation is excluded from a zone (of width δ) surrounding each pre-existing growing crystal. (Figs. 1d and 1e) These conditions may apply to the case of vapor deposition which is rate limited by surface diffusion. In this case there will be a region around a growing crystal in which the concentration of adsorbed atoms is sufficiently depressed to effectively prevent nucleation. We assume that the width of the nucleation-exclusion zone, δ, remains constant as the crystals grow. This is illustrated in Figure 3. The logical extreme of these cases, in which the nucleation-exclusion zone is very large compared to the growth rate, results in the structure of the Voronoi polygons of the centers of a random sequential packing of non-overlapping disks. (Fig. 1f) Studies of this structure has also been reported in connection with problems outside materials science. [21]

All these different structures have distinct geometric and topological properties. In previous work [10], we have reported distributions of grain area, number of sides per grain, boundary segment length, and triple point angles. These models are the two-dimensional analog of the three-dimensional models of nucleation and growth to impingement reported by Mahin *et al.*[22] and by Saetre *et al.*[23].

Effect of Different Growth-Rate Relationships

The models given above all assume that the growth rate is constant with time for each grain, and identical for all grains. In many physical systems this is not necessarily the case. In this section we consider models in which the growth rate of each grain varies according to its radius. This relationship is assumed to be the same for all grains. These models also assume that the nucleation rate per unit of untransformed area remains constant (continuous nucleation).

We have chosen to model a set of simple relationships in which the growth rate is proportional to a power of the radius, r, as given in Table 1: r^{-3}, r^{-2}, r^{-1}, r^0, and r. Each relationship results in a simple expression for grain radius as a function of the time since its nucleation, t, with the radii proportional to $t^{1/4}$, $t^{1/3}$, $t^{1/2}$, t, and exp(t), respectively. The structures produced by these growth rate relationships are shown in Figure 6.

The most striking difference between different growth-rate relationships is in the shape of the boundaries formed at impingement. When two neighboring grains nucleate simultaneously, and grow together at an identical rate, they impinge along a straight line which is the perpendicular bisector of the line joing the nucleation points. This is true regardless of how the two identical growth rates vary with time. If the two grains nucleate at different times, then the

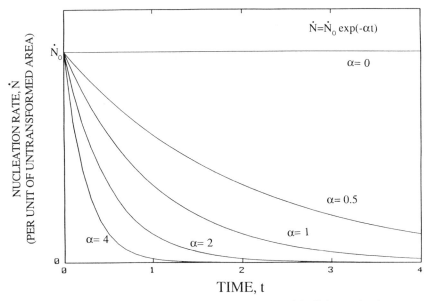

Figure 2. Nucleation rates as a function of time for cases of declining nucleation rates.

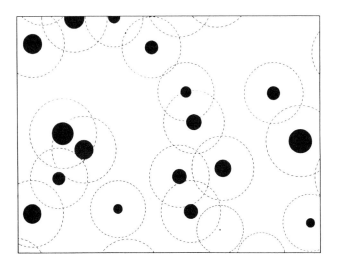

Figure 3. Schematic illustration of the nucleation-exclusion zone model in which the growing crystals are surrounded by an exclusion zone of constant width.

boundary will usually not be a straight line.   Figure 4 shows the impingement of two neighboring grains for different growth-rate relationships, with circles showing the grains at various equally spaced times. Figure 5 shows the family of curves produced by various differences in nucleation time, for each growth rate relationship given in Table I. As mentioned in the previous section, if the growth rate is constant, the boundary will be a hyperbola. If the growth rates decrease with time so that the radius is proportional to $t^{1/2}$, then the boundary is a straight line. If the growth rate decreases more quickly with time ($t^{1/3}$ or $t^{1/4}$), then the boundary will have a region with curvature concave towards the first grain to nucleate, and will eventually asymptotically approach the perpendicular bisector of the line between the nucleation points. If on the other hand, the growth rate increases steadily with radius, the first grain to nucleate will eventually surround and enclose any grain which nucleates at a later time, if it is not otherwise obstructed. When the radius increases exponentially with time, the boundaries are arcs of circles.

The different growth rate relationships allow new topological possibilties. When the radius increases as $t^{1/2}$, and all the boundaries are straight lines, it is impossible to have a grain with less than three sides. When the growth rate is constant, and all the boundaries are hyperbolae, then two-sided grains or lenses become possible. When the growth rate increases with time, it is possible to have completely enclosed or "one-sided" grains. This implies that the area of one grain need not be connected. We effectively assign each point on the plane to the grain which would be the first to reach that point, allowing the grains to hypothetically grow through each other. Since the first grain to nucleate will enclose its smaller neighbors, it can reappear behind a set of smaller neighbors that would block its growth. Although this behavior is not physically realistic, we have included it in our reported results on areas and number of sides.   It is also possible to have disconnected grains when the growth rate decreases with radius faster than $1/r$, when a small grain is nucleated between two large grains just before they meet. For radius proportional to $t^{1/3}$, this is very unlikely; it did not occur once in our sample of about 10,000 grains.

When the growth rate is proportional to radius, we need to introduce an initial radius, r_0, at the time of nucleation: $r(t) = r_0 \exp(Gt)$. Because this initial radius is required, we have chosen to exclude any nucleation that occurs within a distance of r_0 of any pre-existing grain. Different choices of r_0 lead to different structures. These different structures are determined by the relationship among r_0, the constant G (units (time)$^{-1}$) and the nucleation rate $\dot{N}$ (units (time)$^{-1}$ (area)$^{-1}$), which may be expressed in terms of the dimensionless expression: (r_0^2 $\dot{N}/G$). We have chosen to use values of this expression of 0.01, 0.04, and 0.016. As this

Table I. Growth Rate Formulae

	$r(t)$	$\dfrac{dr}{dt}(r)$	$\dfrac{dV}{dt}(r)$	Boundary Curve
1.	$(Gt)^{1/4}$	$\dfrac{G}{4\,r^3}$	$\dfrac{3HG}{r}$	Unnamed $\quad x^2 + y^2 - \dfrac{G\,\Delta t}{8\,L\,x} + L^2 = 0$
2.	$(Gt)^{1/3}$	$\dfrac{G}{3\,r^2}$	$H\,G$	Unnamed $\quad [(x+L)^2+y^2]^{3/2} - [(x-L)^2+y^2]^{3/2} = G\,\Delta t$
3.	$(Gt)^{1/2}$	$\dfrac{G}{2\,r}$	$\dfrac{3HGr}{2}$	Straight Line $\quad x = \dfrac{G\,\Delta t}{4\,L}$
4.	Gt	G	$3\,H\,G\,r^2$	Hyperbola $\quad x^2\left[\dfrac{4\,L^2}{G^2(\Delta t)^2} - 1\right] - y^2 = L^2 - \dfrac{G^2(\Delta t)^2}{4}$
5.	$r_0 \exp[Gt]$	$G\,r$	$3\,H\,G\,r^3$	Circle $\quad y^2 + x^2 + 2\,x\,L\left[\dfrac{1 + e^{2G\Delta t}}{1 - e^{2G\Delta t}}\right] + L^2 = 0$

t = Time since nucleation. Δt = Difference in nucleation times.
V = Grain Volume: $V = H\,r^3$.   G = Growth Rate Constant. (different units for each case.)
r = Grain Radius.   H = Geometric Constant; Hemispherical caps $\Rightarrow H = 2\pi/3$.
Boundary curve formulae are given for two nuclei at $x = \pm L$, $y = 0$.

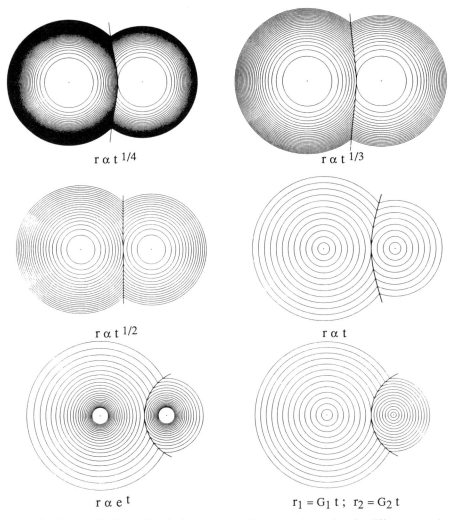

$r \propto t^{1/4}$

$r \propto t^{1/3}$

$r \propto t^{1/2}$

$r \propto t$

$r \propto e^t$

$r_1 = G_1 t \; ; \; r_2 = G_2 t$

Figure 4 - Boundaries formed by the impingement of two growing grains, for different growth-rate relationships. The circles show the size the grains at various equally-spaced times.

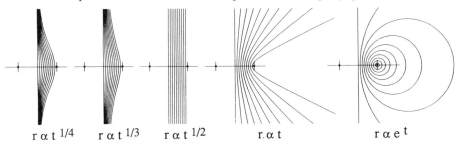

$r \propto t^{1/4}$ $r \propto t^{1/3}$ $r \propto t^{1/2}$ $r. \propto t$ $r \propto e^t$

Figure 5 - Families of the possible boundary curves, for different growth-rate relationships.

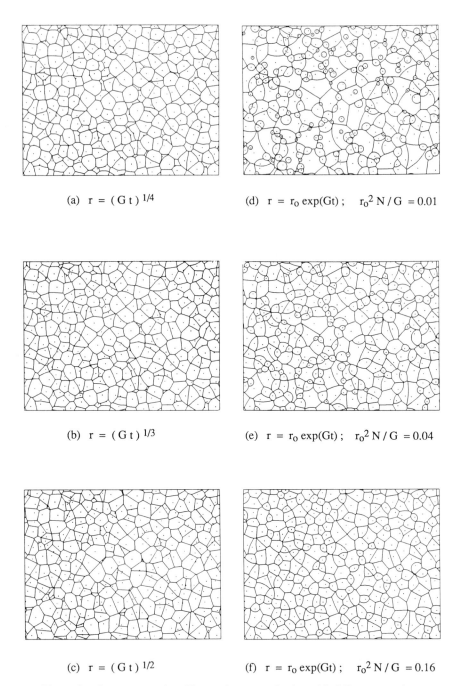

(a) r = (G t)$^{1/4}$

(d) r = r_0 exp(Gt) ; r_0^2 N / G = 0.01

(b) r = (G t)$^{1/3}$

(e) r = r_0 exp(Gt) ; r_0^2 N / G = 0.04

(c) r = (G t)$^{1/2}$

(f) r = r_0 exp(Gt) ; r_0^2 N / G = 0.16

Figure 6 - Structures produced by continuous nucleation with different growth-rate
relationships.

value increases, the ratio of r_o to the average grain area increases, and the structure approaches those with a significant nucleation-exclusion zone, as presented in the previous section.

Although the effects of variable growth-rates have not been widely discussed in the materials science literature, some structures similar to those we have calculated have been discussed in other contexts. Boots [24,25] discussed the applications of weighted Theissen polygons in geography. Different weighting schemes are analogous to different growth-rate relationships. A division of space that is equivalent to the $t^{1/2}$ growth-rate relationship has been used to assign volumes to atoms of different sizes. [26,27] There has also been some analysis of efficient algorithms for calculating the structures produced by different weighting schemes or different growth-rate relationships. [28,29]

In previous work [12] we have reported distributions of the number of sides per grain and the grain area for the various growth-rate relationships. In future work we expect to report more fully on the geometric and topological properties of these structures.

Modelling of Normal Grain Growth

After nucleation and growth to impingement, the microstructures of thin films often continue to evolve by grain boundary migration. If the effects of the film's free surface can be ignored, and the grains are large compared to the film thickness, two-dimensional grain growth may result. We have developed a computer simulation of such grain growth. Boundary migration is assumed to proceed at a velocity proportional the local curvature, κ: $v = \mu \kappa$, where μ is a mobility constant which includes the boundary surface energy. The migration is also assumed to produce angles of 120° at the triple points where three grains meet, which are the equilibrium angles if all boundaries have the same energy. Our simulation technique differs from those used by other simulations of two-dimensional grain growth. [30-37] We have chosen to model boundary migration by following the boundary itself as an array of moving points. (Figure 7) In each time increment the points are moved perpendicular to the boundary line by a distance proportional to local curvature. After each iteration of boundary motion, the triple point positions are adjusted until the tangents to the three boundary segments at the triple point·meet to form angles of 120°. Details of the algorithms used are described elsewhere. [11,38] Our technique has the advantage of increased accuracy, especially during the early stages of the evolution. We have found that the growth or shrinkage rate in area of each grain depends only on its number of sides, within reasonable numerical accuracy, as predicted theoretically. [11]

If sufficient grain boundary migration occurs, the smaller grains will shrink to annihilation, and the average grain area will grow. Under these conditions we find that the initial evolution depends strongly on the initial microstructure. After sufficient time, however, we find that different initial structures converge towards similar structures for which the average grain size (square root of average grain area) increases as the square root of time. An example of the evolution of the microstructures is shown in Figure 8. During this evolution time is conveniently measured in dimensionless units as $\tau = t\,\mu/A_o$, where A_o is the initial average area. Preliminary results indicate that the normalized distribution of grain sizes does not change after the steady-state structures are reached. Figure 9 shows the evolution of the grain size distributions for three different initial structures. This behavior may be considered to be normal grain growth in two dimensions.

Figure 7 - Schematic illustration of the simulation techniques for modelling boundary migration and triple point motion.

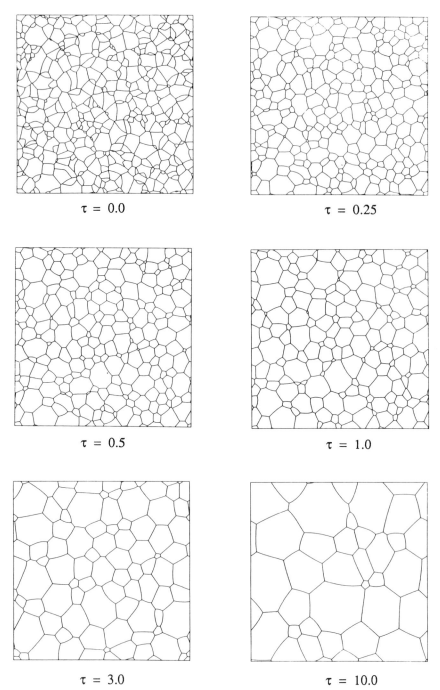

$\tau = 0.0$

$\tau = 0.25$

$\tau = 0.5$

$\tau = 1.0$

$\tau = 3.0$

$\tau = 10.0$

Figure 8 - Evolution of the Continuous Nucleation structure at various normalized times: $\tau = \mu t / A_0$. The initial structure contains approximately 400 grains of unit average area.

88

We have analyzed the grain growth kinetics of our simulation in terms of the commonly used relationship between the average effective diameter (square root of grain area), D, and time, t:

$$D^m - D_0{}^m = B t \qquad (1)$$

where D_0 is the initial diameter at $t = 0$, B is a constant, and m is a kinetic exponent. When $D \gg D_0$, this becomes $D \approx (Bt)^{1/m} = (Bt)^n$. An analysis for interface limited particle coarsening (which is assumed to be analogous to grain growth) suggests that n should be 1/2, if the evolution is "self-similar", such that the structure remains topologically and geometrically equivalent, distinguished only by a length scaling factor such as D, and if the rate of evolution is proportional to an average curvature. [39-42]

For our analysis of the grain growth kinetics, we assume $m = 2$, and re-express equation 1 in terms of the average grain area, $A = D^2$, and the initial average area, $A_0 = D_0{}^2$:

$$\frac{A - A_0}{A_0} = \frac{B}{\mu}\, \tau \;. \qquad (2)$$

Figure 10a shows that this relationship is not obeyed during the early stages of growth, but is approached at later times. This is to be expected because the initial structures are clearly not self-similar to the later structures. The continuous nucleation structure contains many small grains, with few sides, which are annihilated rapidly, causing a rapid initial increase in grain size. In contrast, the structure for the packing of non-overlapping disks has a very narrow distribution of areas, with almost all grains having five, six or seven sides, and therefore shows a long transient in which no grains are annihilated, and average area does not change. It is possible to exclude the initial transient behavior by taking the structure at some later time as the initial structure. In Figure 10b, we have started at $\tau = 2$, or $t = t_0 = 2\, A_0/\mu$, and have defined a new dimensionless time, $\tau^* = (t - t_0)\, \mu/A_0{}^*$, where $A_0{}^*$ is the average area at $t = t_0$. The results clearly demonstrate, within the limits of our statistical sample size, that equation 2 applies, and that the evolution is characterized by $m = 2$ or $n = 1/2$.

The steady-state structures produced by our simulation appear to be qualitatively similar to planar sections through three-dimensional grain growth microstructures. It is improper, however, to make a detailed comparison of our two-dimensional model, or any other two-dimensional model, with the planar section through a three-dimensional microstructure, because the topology of the two cases is inherently different. For example, in our model, we have not allowed the nucleation of new grains after the grain growth has started. In a three-dimensional structure, a planar section will have new grains appear on it as nearby, growing grains impinge on that plane. In addition, the distribution of triple-point angles on a planar section is not identical to the dihedral angles at grain edges in three-dimensions. Because normal, steady-state grain growth is not commonly observed in thin films, as discussed in the next section, it is therefore difficult to compare our steady-state microstructures with experimental micro-structures.

Modelling of Abnormal Grain Growth

It is experimentally observed that normal grain growth in thin films generally ceases when the average grain diameter is two or three times the thickness. [43] When further grain growth occurs, it is by the abnormal or secondary growth of a few of the grains which outstrip and consume their neighbors. [13] There may be several factors which contribute to this behavior. As the grains grow larger than the film thickness, there is shift from boundary curvature in two directions to curvature in only one direction. This geometric effect should slow down grain growth, but should not be sufficient to account for the observed stagnation of grain growth. One cause of this stagnation may be grain boundary grooving at free surfaces, which places a drag on further boundary migration. [44] The boundaries may also be pinned by precipitate particles. When the grains become much larger than the film thickness, the area of the free surface of the film exceeds the area of the grain boundaries. If the anisotropy of free surface energy is sufficiently high, the driving force of minimization of free-surface energy should become important. Grains oriented for low free-surface energy would be favored. Even if the

89

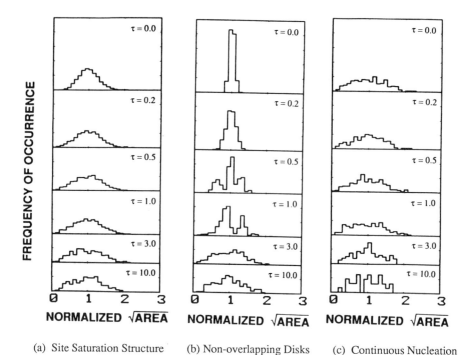

(a) Site Saturation Structure   (b) Non-overlapping Disks   (c) Continuous Nucleation

Figure 9 - Evolution of effective grain diameter distributions for three structures during grain growth. Each distribution is normalized to the average area at the particular normalized time, τ.

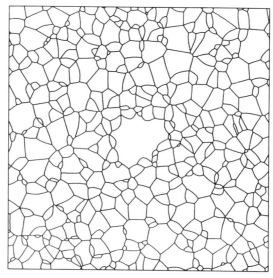

Figure 11 - Initial structure used for abnormal grain growth simulation. The structure was formed by continuous nucleation and growth to impingement, with the central grain started 1.5 time units before nucleation of the other grains began.

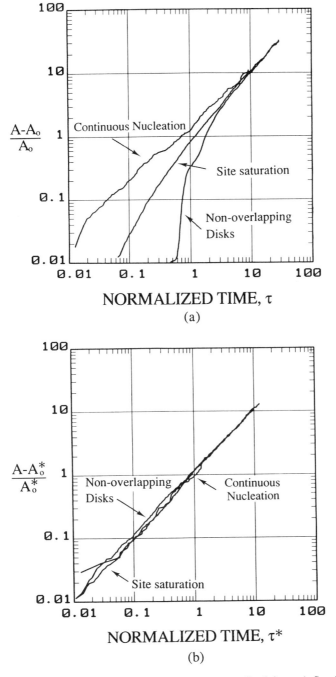

Figure 10 - Normalized average grain area versus normalized time. a) Starting from the initial configurations. b) Starting from the structure present at $\tau = 2.0$.

anisotropy of free-surface energy is not high enough to entirely dominate boundary energy minimization, it could provide a force sufficient to overcome boundary pinning for some grains.

In order to understand more fully the conditions which lead to abnormal or secondary grain growth, we have begun to apply our grain growth simulation to the problem. When abnormal grain growth does occur, there are usually many different abnormal or secondary grains in a film, which grow until they impinge on each other. The final grain size after abnormal grain growth is therefore determined by the density of abnormal grains. That is, the final structure is determined by the detailed conditions which specify how many of the grains in the initial structure will become abnormal grains. (The initial structure may be that which results from limited normal grain growth.) Our goal is therefore to determine exactly what conditions will lead grains to grow abnormally. To achieve this we must explicitly consider the initial microstructure from which the abnormal grains will emerge. Our previous modelling provides the most realistic initial microstructures available.

We may imagine three different ways to achieve abnormal grain growth. First, the abnormal growth might result for a particular grain if it is much larger than the other grains in the microstructure. Second, the abnormal growth might result because the abnormal grain is surrounded by grain boundaries which have a higher mobility than the other grain boundaries in the structure. Third, the abnormal growth might result from the existence of a driving force in addition to that provided by the grain boundary energy, which drives the boundaries surrounding the abnormal grain outward. Under some conditions, the effect of anisotropy in free-surface energy might represent an example of this third case. In an experimental system, there may components of all three cases.

The first case, which we may call geometrically-driven abnormal growth, was proposed by Hillert [41]. A reconsideration of how Hillert's model for grain growth applies to abnormally large grains [45] has shown that his model does not predict geometrically-driven abnormal growth. If the only driving force for grain growth is a constant, isotropic grain boundary energy, and if the mobility is the same for all boundaries, then although an extra large grain will grow faster than its neighbors, it will not grow relatively faster than the average normal grain size. That is, the ratio of the size of the abnormal grain to the average normal grain size will decrease until the abnormal grain has rejoined the normal size distribution. This prediction has been supported by the abnormal grain growth simulations of Srolovitz et al. [46]. If the initial structure does not contain any abnormally large grains, we do not expect them to develop from geometrical considerations alone.

The second and third cases have inherent differences. For a grain that is already abnormally large, both cases would favor abnormal growth. For normal grains in the initial microstructure, this is not always true. If a particular grain has boundaries with a higher mobility (case two), then it might *collapse* faster than it otherwise would, rather grow faster than it otherwise would. If a particular grain has an additional driving force for growth (case three), then it will either *collapse more slowly* or *grow faster* than it otherwise would. The simulation of Srolovitz et al. [46] considered case three in which a fraction of the initial grains were favored. More recent work of the Anderson, Srolovitz et al. [47,48] has considered the second case. In this work we also present results for the second case, in which the boundaries surrounding one grain are given a higher mobility than the other boundaries. To insure that this grain does grow abnormally, we must start with a large grain which would be growing even without enhanced mobility.

For illustration of abnormal grain growth, we have chosen an initial structure containing a central grain which is larger than any grains in the normal distribution, as shown in Figure 11. This structure results continuous nucleation and growth to impingement with the central grain nucleated 1.5 time units before nucleation of the other grains began. (The growth rate, 0.82467 length per time units, is chosen so that the average normal area is unity.) In this example, the initial abnormal grain area is 11.6. We have used three different values for the mobility constant (μ_{ab}) for the boundaries that surround the central, abnormal grain, which are 1, 2 and 4 times the mobility constant used for the normal grain boundaries. The first case (mobility ratio = 1) is just a condition to test for geometrically-driven abnormal growth, without enhanced mobility. Figure 12 shows the resulting structures for various normalized times, $\tau = \mu t / A_0$, where μ is for normal boundaries. In all cases, the central grain gains area at a faster rate than the increase

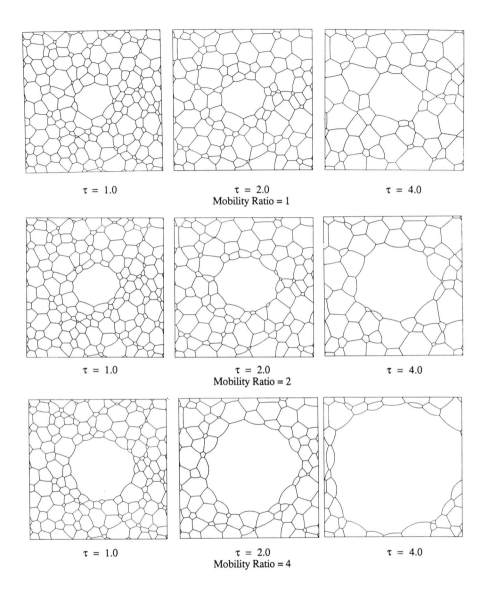

$\tau = 1.0$ $\tau = 2.0$ $\tau = 4.0$

Mobility Ratio = 1

$\tau = 1.0$ $\tau = 2.0$ $\tau = 4.0$

Mobility Ratio = 2

$\tau = 1.0$ $\tau = 2.0$ $\tau = 4.0$

Mobility Ratio = 4

Figure 12 - Abnormal grain growth structures at various normalized times: $\tau = \mu t/A_o$, for three different ratios of abnormal to normal grain boundary mobility.

in average grain area, as shown in Figure 13. It is more instructive, however, to consider the *relative* growth rates, by measuring the ratio of the abnormal grain size to the average normal grain size. This is shown in Figure 14, which shows ratios of effective diameters (square roots of area). For the first example, mobility ratio = 1, the abnormal grain grows relatively slower than the average normal grain size, verifying theoretical prediction [45] and previous simulation [46]. It is interesting to note that the ratio appears to asymptotically approach a value of about 2.5, whereas Hillert's analysis predicts that the ratio should approach 2.0. Further simulation runs are required before this difference can be meaningfully verified.

For the case where the boundary mobility ratio = 4, it is clear that the abnormal grain continues to grow relatively faster than the normal grain size throughout the simulation. For the intermediate case, mobility ratio = 2, the abnormal grain has only slightly faster relative growth. Although these preliminary simulations do not accurately determine the relationship between the boundary mobility ratio and the relative growth rates, they clearly demonstrate the trend. In future work we will extend the simulation effort so that we may more accurately understand how the variables which control abnormal grain growth operate to determine the density of abnormal grains and the resulting microstructure.

Conclusions

For a complete understanding of the development of microstructure of a polycrystalline thin film, there are a variety of processes which must be considered. Among these are the nucleation of crystals, the growth of crystals to impingement, and the migration of boundaries after impingement. Of the many possible combinations of nucleation conditions and growth relationships, we have chosen to model a few simple nucleation conditions with a constant growth rate, and a few simple growth-rate relationships with a constant nucleation rate. These different cases produce widely different geometries and topologies. They also provide an indication of how more complex combinations of nucleation conditions and growth-rate relationships will effect the microstructure. We have also undertaken the simulation of grain boundary migration and grain growth with a technique which is well suited for demonstrating the important effects of the initial microstructure. The simulation has also been applied to the case of normal, steady-state grain growth in two dimensions, and to a few cases of abnormal or secondary grain growth. Through modelling of this type, we hope to build an understanding of how the conditions of film formation control the resulting microstructures.

Acknowledgements

This work was supported by the National Science Foundation, grant number DMR-8506030. The computer program for the simulation of grain growth was created by Junho Whang and Caroline Howe, with amendments by D. Walton. We are also grateful to D. Walton, E. Gewirtz, and D. Call for preparation of illustrations.

References

1. Y.Pauleau, "Interconnect Materials for VLSI Circuits," Solid State Technology, Feb. 1987, 61-67
2. S. D. S. Malhi et al., "Characteristics and Three-Dimensional Integration of MOSFET's in Small Grain LPCVD Polycrystalline Silicon", IEEE Trans. on Electron Devices ED-32 (1985) 258-281
3. J.-W. Lee, B.G. Demczyk, K.R. Mountfield and D.E. Laughlin, "Transmission electron microscopy of Co-Cr films for magnetic recording", J. Appl. Physics 61 (1987) 3813-3815 (see also other articles in this issue)
4. R. T. Howe, Annals of Biomedical Engineering 14 (1986) 187
5. G. L. Schnable, W. Kern and R. B. Comizzoli, "Passivation Coatings on Silicon Devices", J. Electrochem. Soc. 122 (1975) 1092-1104
6. F. M. d'Heurle and P. S. Ho, Chapter 8 in Thin Films--Interdiffusion and Reactions, edited by J. Poate, K.-N. Tu and J.W. Mayer, (Wiley Interscience, 1978), 243

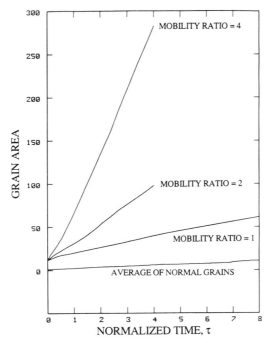

Figure 13 - Areas of abnormal grains as a function of normalized times for three different ratios of abnormal to normal grain boundary mobility: 1, 2, and 4. Also shown is the average area for normal grain growth from the continuous nucleation initial structure.

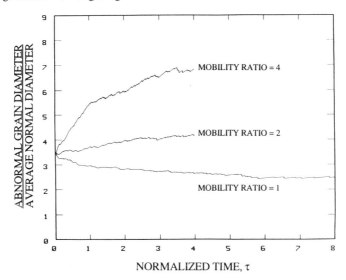

Figure 14 - Evolution of the ratio of the abnormal grain size (square root of area) to the average normal grain size (square root of average area), for three different ratios of abnormal to normal grain boundary mobility: 1, 2, and 4.

95

7. M. J. Attardo, R. Rutledge, and R. C. Jack, "Statistical Metallurgical Model for Electromigration Failure in Aluminum Thin-Film Conductors", J. Appl. Physics 42 (1971) 4343-4349

8. F. M. d'Heurle and I. Ames, "Electromigration in Single-Crystal Aluminum Films", Applied Physics Letters 16 (1970) 80-81

9. S. Vaidya, T. T. Sheng, and A. K. Sinha, "Linewidth dependence of electromigration in evaporated Al - 0.5 % Cu", Applied Physics Letters 36 (1980) 464-466

10. H.J. Frost and C.V. Thompson, "The Effect of Nucleation Conditions on the Topology and Geometry of Two-Dimensional Grain Structures," Acta Metallurgica 35 (1985) 529-540

11. H.J. Frost, C.V. Thompson, C.L. Howe and J. Whang, "A Two-Dimensional Computer Simulation of Capillarity-Driven Grain Growth: Preliminary Results", Scripta Metallurgica 22 (1987) 65-70

12. H.J. Frost and C.V. Thompson, "Development of Microstructure in Thin Films", Proceedings of SPIE - The International Society for Optical Engineering, Volume 821, Modeling of Optical Thin Films, M.R. Jacobson chair/editor, Aug. 16-17, 1987, 77-87

13. C.V. Thompson, "Observations of Grain Growth in Thin Films", This Symposium

14. A. Getis and B.N. Boots, Models of Spatial Processes, (Cambridge, U.K.: Cambridge University Press, 1979)

15. W.A. Johnson and R.F. Mehl, "Reaction Kinetics in Processes of Nucleation and Growth," Trans AIME 135 (1939) 416-458

16. U.R. Evans, "The Laws of Expanding Circles and Spheres in Relation to the Lateral Growth of Surface Films and the Grain-Size of Metals", Trans. Faraday Society 41 (1945) 365-374

17. J.L. Meijering, "Interface Area, Edge Length, and Number of Vertices in Crystal Aggregates with Random Nucleation", Philips Res. Rep. 8 (1953) 270-290

18. E.N. Gilbert, "Random Subdivisions of Space into Crystals", Annals of Mathematical Statistics 33 (1962) 958-972

19. K-N. Tu, D. A. Smith and B.Z. Weiss, "Hyperbolic Grain Boundaries", Phys. Rev. B 36 (1987) 8948-8950

20. D.A. Smith, K-N. Tu and B.Z. Weiss, "Crystallization of Amorphous $CoSi_2$", This Symposium

21. D. Weaire and N. Rivier, "Soap, Cells and Statistics -- Random Patterns in Two Dimensions", Contemp. Physics 25 (1984) 59-99

22. K.W. Mahin, K. Hanson and J.W. Morris, Jr., "Comparative Analysis of the Cellular and Johnson-Mehl Microstructures Through Computer Simulation," Acta Met. 28 (1980) 443-453

23. T.O. Saetre, O. Hunderi and E. Nes, "Computer Simulation of Primary Recrystallisation Microstructures: The Effects of Nucleation and Growth Kinetics," Acta Met. 34 (1986) 981-987

24. B.N. Boots, "Weighting Thiessen Polygons", Economic Geography 56 (1980) 248-257

25. B.N. Boots, "Modifying Thiessen Polygons", The Canadian Geographer 31(2) (1987) 160-169

26. B.J. Gellatly and J.L. Finney, "Characterisation of Models of Multicomponent Amorphous Metals: The Radical Alternative to the Voronoi Polyhedron", J. Non-Crystalline Solids 50 (1982) 313-329

27. W. Fischer and E. Koch, "Geometrical Packing Analysis of Molecular Compounds", Zeitschrift fur Kristallographie 150 (1979) 245-260

28. P.F. Ash and E.D. Bolker, "Generalized Dirichlet Tessellations", Geometriae Dedicata 20 (1986) 209-243

29. F. Aurenhammer and H. Edelsbrunner, "An Optimal Algorithm for Constructing the Weighted Voronoi Diagram in the Plane", Pattern Recognition 17 (2) (1984) 251-257

30. D. Weaire and J.P. Kermode, "The Evolution of the Structure of a Two-Dimensional Soap Froth", Phil. Mag. B 47 (1983) L29-L31; "Computer Simulation of a Two-Dimensional Soap Froth, I. Method and Motivation; II. Analysis of Results", 48 (1983) 245-259; 50 (1984) 379-395

31. J. Wejchert, D. Weaire and J. P. Kermode, Phil. Mag. B 53 (1986) 15

32. M.P. Anderson, D.J. Srolovitz, G.S.Grest, and P.S. Sahni, "Computer Simulation of Grain Growth--I. Kinetics", Acta Met. 32 (1984) 783-791

33. D.J. Srolovitz, M.P. Anderson, P.S. Sahni, and G.S. Grest, "Computer Simulation of Grain Growth--II. Grain Size Distribution, Topology, and Local Dynamics", Acta Met. 32 (1984) 793-802

34. E.A. Ceppi and O.B. Nasello, "Computer Simulation of Bidimensional Grain Growth", Scripta Met. 18 (1984) 1221-1225

35. A. Soares, A.C. Ferro and M.A. Fortes, "Computer Simulation of Grain Growth in a Bidimensional Polycrystal", Scripta Met. 19 (1985) 1491-1496

36. S. Yabushita, N. Hatta, S. Kikuchi, and J. Kokado, "Simulation of Grain-Growth in the Presence of Second Phase Particles", Scripta Met. 19 (1985) 853-857

37. V.E. Fradkov, L.S. Shvindlerman, and D.G. Udler, "Computer Simulation of Grain Growth in Two Dimensions", Scripta Met. 19 (1985) 1285-1290

38. C.L. Howe, Computer Simulation of Grain Growth in Two Dimensions, M.E. Thesis, Dartmouth College (1987)

39. C. Wagner, "Theorie der Alterung von Neiderschlägen durch Umlösen", Zeitschrift für Elektrochemie 65 (1961) 581-591

40. I.M. Lifshitz and V.V. Slyozov, Zh. Eksp. Teor. Fiz. 35 (1958) 479

41. M. Hillert, "On the Theory of Normal and Abnormal Grain Growth", Acta Met. 13 (1965) 227-238

42. W.W. Mullins, "The Statistical Self-Similarity Hypothesis in Grain Growth and Particle Coarsening", J. Appl. Phys. 59 (1986) 1341-1349

43. P.A. Beck, M.L. Holtzworth and P.R. Sperry, Trans. AIME 180 (1949) 163

44. W.W. Mullins, "The Effect of Thermal Grooving on Grain Boundary Motion", Acta Met. 6 (1958) 414-427 (1958)

45. C.V. Thompson, H.J. Frost and F. Spaepen, "The Relative Rates of Secondary and Normal Grain Growth", Acta Met. 35 (1987) 887-890

46. D.J. Srolovitz, G.S. Grest and M.P. Anderson, "Computer Simulation of Grain Growth--V. Abnormal Grain Growth", Acta Met. 33 (1985) 2233-2247

47. M.P. Anderson, G.S. Grest and D.J. Srolovitz, "Computer simulation of grain growth", Third International Conference on Progress in Microstructure, Aachen, West Germany, May 4-8, 1987

48. A.D. Rollett, D.J. Srolovitz and M.P. Anderson, "Simulation and Theory of Abnormal Grain Growth - Anisotropic Grain Boundary Energies and Mobilities", submitted to Acta Metallurgica (1988)

COMPUTER SIMULATION OF ZONE II MICROSTRUCTURAL DEVELOPMENT

IN THIN FILM MATERIALS

S. Ling and M. P. Anderson

Exxon Research and Engineering Co., Annandale, N.J. 08801

Abstract

A computer simulation procedure based on the Monte Carlo algorithm has been developed to study the thin film microstructure evolution in the Zone II regime. This is accomplished by letting the grain boundary movement on the film surface couple with the microstructure in the interior of the film. The grains when viewed in cross-section are found to reach a limiting size at which point all grain boundary motion ceases despite the presence of capillarity driving force. This results in a columnar morphology for the thin film. The cross-sectional size distribution funcion is found to be approximately log-normal in shape. The introduction of anisotropic solid-vapor interfacial energy is found to induce abnormal grain growth at the expense of the normal grain which has a relatively higher solid-vapor energy. An analytical model is proposed in which the formation of the columnar structure is explained in terms of the grain boundary drag at the growing interface due to sub-surface boundaries.

Microstructural Science for Thin Film
Metallizations in Electronics Applications
Edited by J. Sanchez, D.A. Smith and N. DeLanerolle
The Minerals, Metals & Materials Society, 1988

Introduction

A major factor affecting the physical properties of thin films is the microstructure. Over the last two decades considerable experimental effort has been expended to classify thin film morphologies [1,2,3], and it was found that the microstructure of a thin film depends on many factors. Take, for example, the microstructure of a thin film formed by the physical vapor-deposition process. This depends upon, among other things, the deposition rate, the substrate surface roughness, as well as the substrate temperature. The morphology of thin films formed in this fashion can in general be described by the well known three Zone description, in which, depending on the substrate temperature, the film acquires one of three distinct types of morphology [1,2]. At low substrate temperature, we have Zone I, where the deposited atoms do not have sufficient thermal energy to diffuse away to form a continuous film, resulting in a porous structure which usually has void regions in between. At intermediate temperature, in the Zone II regime, surface diffusion dominates which results in a film consisting of columnar grains with fully dense boundaries. At higher temperature up to the melting point we have Zone III, in which the temperature is high enough to permit substantial bulk diffusion such that recrystallization and grain growth may take place, resulting in a microstructure consisting of equiaxed grains.

Despite the importance of the physical structure of thin films, only a qualitative understanding of the mechanisms underlying the formation of the microstructures in these Zones has been developed. One important model is the ballistic aggregation studies of Meakin et al., which predicted the porous microstructure of Zone I [4]. Recently Srolovitz also proposed an analytical model to explain the formation of columnar grain structure in Zone II [5].

In the past few years there has been a development which has attracted a lot of attention. This is the growing activity in the modelling of grain growth in two dimensional systems using various computer simulation techniques. Several different approaches have been adopted by various authors. Weaire et al. have reported a model for the evolution of a two dimensional soap froth [6,7]. Frost et al. developed a procedure in which the grain boundaries are moved while maintaining 120° angle at the triple points [8]. A different approach developed by Anderson et al. utilized a digitized microstructure which is allowed to have temporal evolution with a Monte Carlo procedure [9,10,11,12]. Various attempts have been made in applying the findings of these works to the study of thin films despite the fact that a thin film has a third dimension, namely its thickness, and is not a true two dimensional system. It is the effect of the film thickness on the microstructure that we want to incorporate into our model in this study.

In this paper we attempt to simulate the temporal evolution of the Zone II thin film microstructure. The simulation procedure we used is based on the Monte Carlo procedure developed by Anderson et al., in which the driving force of the grain boundary movement is the capillarity. In order to take into account the three dimensional nature of the thin film, the procedure is modified to include the effect of the underlying microstructure as well as that of the solid-vapor interfacial energy on the thin film growth. To simplify our study we have neglected the effect of the surface grooving. The substrate temperature is assumed to be low enough that grain boundary migration is allowed only in a thin layer at the film-vapor interface. This assumptions may be the essential experimental conditions for the formation of the Zone II grain morphology [2]. The other topic we want to address in our study is the abnormal grain growth in thin film, which is incorporated

100

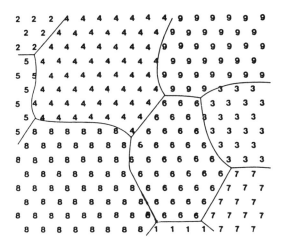

Fig. 1   An example of a microstructure being digitized in which a grain
structure is being mapped onto a triangular lattice.

into our model by allowing certain favored grain orientations to have lower
solid-vapor interfacial energies.

Simulation Procedure

The simulation is performed with a digitized microstructure, which is
obtained by mapping a continuum microstructure onto a two dimensional lattice. To each site is assigned a number, S_i, between 1 and Q which represents the crystallographic orientation of the grain in which the site is
embedded. A grain boundary segment is thus defined to lie between two sites
of unlike orientations, whilst a site surrounded by sites with the same
orientation is in the grain interior. This is shown schematically in Fig.
1. The growth of a thin film by deposition from the vapor phase is then
simulated by successively adding new layers of the two dimensional lattice.

The energy of the system is specified by associating a positive energy
with sites on the grain boundary, as well as with sites on the solid-vapor
interface, and zero energy for sites in the grain interior, according to

$$H = \sum_i \left(\sum_{nn} -J(\delta_{S_iS_j} - 1) + H^{sv} \right) \qquad (1)$$

where δ_{ij} is the Kronecker delta, J is a positive constant that sets the
scale of the grain boundary energy, and H^{sv} the solid-vapor interfacial
energy. $\sum_{nn}$ sums the first term, which represents the contribution of the
grain boundary energy, over all nearest neighbors (nn) of a site. The
energy of the system is then obtained with $\sum_i$ which sums both the grain
boundary as well as the solid-vapor interface energies over the whole thin
film. The kinetics of boundary motion are simulated by employing a Monte

Carlo technique in which a lattice site from the top layer of the film is selected at random, and a new trial orientation is also chosen at random from one of the other (Q - 1) possible orientations. The change in energy, ΔE, associated with this change is evaluated. If ΔE is less than or equal than zero, the re-orientation is accepted. If ΔE, however, is greater than zero, the re-orientation is accepted with a probablity of exp(-ΔE/kT), where kT is the thermal energy. Boundary migration kinetics specified in this fashion are formerly equivalent to a rate theory description [13]. Since only sites in the top layer are chosen, it is assumed, in accordance with the Zone II criterion, that the microstructure in the thickness of the film is frozen and any evolution of microstructure can take place only in the surface layer of the film (i.e., at the growing interface). Time, in these simulations, is related to the number of re-orientation attempts. N re-orientation attempts are arbitrarily used as the unit of time and is referred to as 1 Monte Carlo Step (MCS), where N is the number of lattice sites. The conversion from MCS to real time thus has an implicit activation energy factor, exp(-W/kT), which corresponds to the atomic jump frequency and is not determined by the simulation.

In this study, the simulations are performed with $Q = 48$, $T \approx 0$. The lattice used was a 40,000 sites square lattice with periodic boundary condition. The site interaction is extended to include up to the seventh nearest neighbors (see equation (1)) in order to eliminate the effect of the discrete lattice on the outcome of the simulation. For the study of normal grain growth in thin film, H^{sv} is set to be equal to J for all surface sites so that no one orientation has any energy advantage over the others. The initial layer of the film has random orientations assigned to all of its lattice sites to simulate the randomly orientated fine grains which are deposited at the film-substrate interface. The microstructure in the surface layer is allowed to evolve for one MCS before a new layer is laid down over it. The freshly deposited layer always assumes a grain structure of the layer immediately beneath it, which is then allowed to evolve for another MCS in the manner described above before the next layer is deposited. For the study of abnormal grain growth in thin film, a microstructure obtained from the normal grain growth study is used as the initial layer in which approximately 2% of the grains are selected at random to be the abnormal grains. It is observed that in abnormal grain growth in thin films the surface energy of a grain depends on both the crystallographic orientation of the grain's surface planes as well as the nature of the atmosphere [14,15,16]. It is conceivable that certain orientations can thus have a much lower surface energy than surrounding grains which gives them an advantage in growth. To simulate this effect, the H^{sv} of the chosen grain is set to be αJ, where $\alpha < 1$. These grains are also relabeled so that each of them has a unique grain orientation. The kinetics of the abnormal grain growth are then simulated in the same manner as that of the normal grain growth.

Results

Normal grain growth in thin film

Thin film growth is simulated with the procedure outlined in the previous section. The resulting morphology after 5420 MCS is shown in Fig. 2, which shows both the cross-sectional top view of the top-most layer as well as the in depth side views of the film. Several features are evident from the top view of the film: First of all the microstructure has uniform grain size in the sense that the largest grains have their radius within three times the size of the mean grain radius. This can be seen clearly in the cross-sectional size distribution shown in Fig. 3 which shows that the

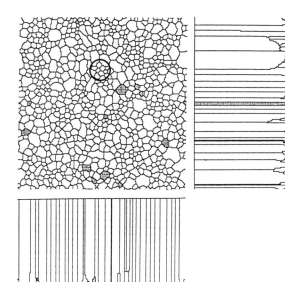

Fig. 2 The microstructure
of the thin film after 5420
MCS of normal grain growth
simulation. Shown in the
figure are the cross-
sectional top view of the
top layer as well as two
in-depth side views of the
film. The circle indicates
some small grains with six
or fewer sides.

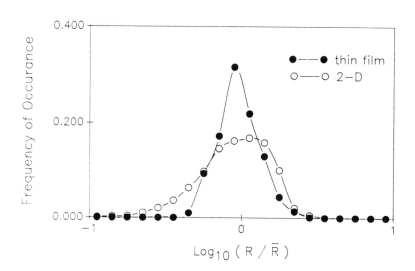

Fig. 3 The plot of frequency of occurance versus the normalized cross-
sectional grain radius, $R/\langle R \rangle$, where $\langle R \rangle$ is the mean grain radius.
The total area under each of the curves is normalized to one. Shown
in this figure are the plots obtained for the case of normal grain
growth in a thin film as well as that of grain growth in a pure two
dimensional system.

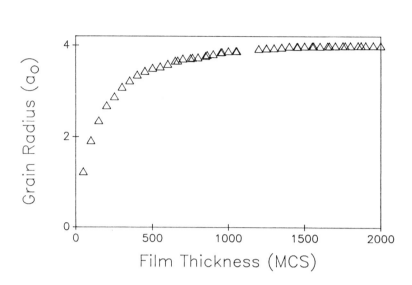

Fig. 4 The plot of mean cross-sectional grain radius versus film thickness, obtained for normal grain growth.

upper cut-off point of the grain size distribution occurs at $R/\langle R \rangle \simeq 3$, where R is the grain radius, and $\langle R \rangle$ the mean grain radius. Secondly the grain boundaries tend to meet at 120^0 angle. Lastly there is no lack of grain boundary curvature, indicating the presence of driving force for further microstructure development. These are all features that are expected of a normal bulk grain growth process. The side views of the film in Fig. 2, on the other hand, revealed some interesting features. It can be seen that near the film-substrate interface the film consists of many fine grains undergoing growth competition. As the film grows away from the interface and the grains become larger, a transition takes place after which the grain boundary motion is all but ceased despite the presence of capillarity driving force still present in the top layer. The grains then continue to grow upwards which results in the columnar morphology shown, with a mean cross-sectional radius of four lattice units. This columnar morphology of thin film shown in Fig. 2 is consistent with that observed in many experiments [17,18].

The cross-sectional grain size distribution function obtained for the surface layer of the film after 5420 MCS is shown in Fig. 3, which also included for comparision purpose a distribution function obtained from a pure two dimensional grain growth study. Both curves peak at $R/\langle R \rangle = 1$, and have the same upper cut-off at $R/\langle R \rangle \simeq 3$. They are, however, different in their symmetry. The two dimensional curve is asymmetric and is skewed at small grain size, whereas the thin film curve is symmetric and appears to be a log-normal curve. A thin film grain size distribution function similar to what we obtained here is observed experimentally elsewhere [19].

The mean cross-sectional grain radius versus the film thickness is plotted in Fig. 4, where the film deposition time is taken as a measure of its thickness. Since we have not yet established a relationship between the time scale of our simulation and that of a real film growth, the film thickness in this figure is given in MCS time units. The plot is scaled in such a way as to show the trend in grain size versus time variation. It can be seen that near the film-substrate interface there are many fine grains. As the film grows away from the interface, the rate of growth of the mean grain radius becomes slower, until it finally hit a saturation value of four lattice spacings at around 1000 MCS away from the interface. A similar trend of thin film grain growth was observed experimentally [18].

Abnormal grain growth in thin film

The microstructure shown in Fig. 2 is used as the initial layer for the abnormal grain growth, with the shaded grains selected as the abnormal grains by assigning them a lower solid-vapor interaction energy, H^{SV}. Fig. 5 shows the temporal evolution of the resulting microstructures with $\alpha = 0.025$. It can be seen that the normal grains are frozen, and the abnormal grains expand into them rapidly until impinging on one another. Once the impingement occurs the abnormal grain boundary motion ceases. This is because none of the abnormal grains has any energy advantage over the other now. There is one exception to this observation, however, which occurs at the place indicated by the small circle in Fig. 5. It can be seen that the very sharp beak shaped corner present at 400 MCS was smoothed out later, even though it is completely surrounded by abnormal grains. This is not observed elsewhere where the grain boundary curvature is gentler. This indicates that even with large grains, with large enough capillarity driving force, microstructure evolution is still possible in thin films.

Fig. 6 shows the top as well as two side views of the $\alpha = 0.3$ thin film microstructure obtained after 1200 MCS, when impingement of the abnormal grains is nearly complete. It again reveals clearly many features of similar to that observed in normal grain growth process shown in Fig. 2: uniform grain size, lots of grain boundary curvature, and 120° grain boundary

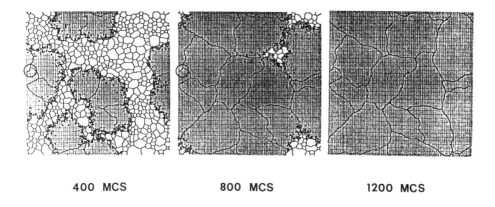

| 400 MCS | 800 MCS | 1200 MCS |

Fig. 5 The microstructures obtained at equal time intervals showing the temporal evolution of abnormal grain growth. The simulation is carried out with $H^{SV} = \alpha J$, where $\alpha = 0.025$. The circles indicate a sharp corner that is present at 400 MCS which gets smoothed out at 800 MCS.

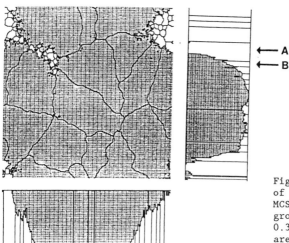

Fig. 6 The microstructure of the thin film after 1200 MCS of abnormal grain growth simulation with α = 0.3. Shown in the figure are the cross-sectional top view of the top layer as well as two in-depth side views of the film. It can be seen from one of the side view that the abnormal grain grows faster into the smaller normal grain labled 'B' than into the larger one labled 'A'.

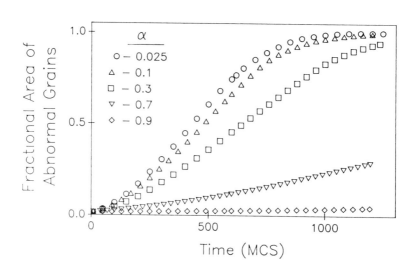

Fig. 7 The plot of area fraction of abnormal grains versus time. The curves are obtained with different solid-vapor interfacial energies, H^{sv}.

intersection. The side views reveal essentially the same things seen in Fig. 5. The abnormal grain grows into the normal grains until impinging with one another, whence their grain boundary movement ceases. The grains then grow upward to form new columnar morphology with larger abnormal grains. It is, however, interesting to notice that the rate of the abnormal grains growing into the normal grains is not constant. Remembering that the thickness of the film is also a measure of the elapsed time, the slope of the boundary that separates an abnormal grain from a normal grain gives a measure of how fast the abnormal grain is expanding into the normal grain. The abnormal grain expands faster when encountering finer normal grains than when encountering larger normal grains. The two grains which are labled 'A' and 'B' respectively, exemplify this observation. This is a result of the fact that more grain boundary surface is eliminated by the expansion of the abnormal grain into finer than larger normal grains.

Fig. 7 shows the fraction of the total cross-sectional area occupied by the abnormal grain as a function of time. Each curve is obtained with a different value of α. The curves display a sigmoidal shape and approach the value of one at infinite time. The curves with the smaller α values rise faster and reach the saturation value of one earlier. For $\alpha < 0.1$, however, the rising rate of the curves seems to have reached a limit and does not change much with continued decrease in α. This results in the curves with $\alpha < 0.1$ overlapping each other.

Discussion

An interesting observation made in our simulation work is that the grain boundary movement ceases once the grains grow to a critical size, leading to the formation of columnar grain morphology. This occurs despite the presence of ample capillarity driving force. Such behavior, however, was not observed in the simulation work carried out by Srolovitz [12], which was based on a procedure similar to our present work except in one vital aspect: Srolovitz did not have the surface microstructure coupled with that in the film interior. This indicates that the 'dragging' effect of the grain boundaries in the thickness of the film is responsible for the observed morphology. We will show in this section that this is indeed the case.

Fig. 8 is a schematic representation of the surface layer of the thin film, which has a constant thickness h. This layer would assume the same grain structure as the underneath layer when freshly deposited. If one of the surface grains, which has a cross-sectional radius R, is to increase its size by δR, all its neighboring grains would shrink and have their radii facing the expanding grain shortened by a corresponding δR. Since the pre-existing microstructure of the underneath layer is frozen (an assumption in our model due to the low substrate temperature), the grain boundaries at the bottom of the surface layer are pinned and can not move. This results in a new morphology characterized by an inverted cone shaped grain as shown in the figure. This new morphology has a net grain boundary area different from that of the initial morphology, which is indicated by the dashed lines in the figure. There are two contributions to this difference in the area: the surface of the grain itself, and the 'flaps' (which are formed by the extension of the surrounding grain boundaries) that have been consumed by the expanding grain. Since the total grain boundary energy is directly proportional to the area, this change in morphology would be energetically favored if a net reduction in grain boundary area is achieved.

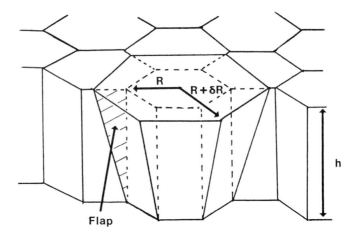

Fig. 8 Schematic representation of a surface layer of material with thickness h that has been freshly deposited on a pre-existing film. The grain indicated by the dashed line has cross-sectional radius R, and is growing to R + δR. Since the grain boundaries at the bottom of the layer are frozen, a new morphology is formed which is characterized by the inverted cone shaped grain. In the process of expanding the grain would also consume the attached triangular shaped 'flaps', one of which is shaded, formed by the extension of the surrounding grain boundaries.

In order to take into account the curvature in the determination of the grain boundary surface area, a construction similar to the one Feltham developed is used [20]. The areas associated with the original and the new morphologies, repectively S_0 and S, are derived in the Appendix. Their ratio is then found to be

$$S/S_0 = (2 + r) \sqrt{(r^2 + t^2)} / 2t - r / 2K(\theta) \qquad (2)$$

where 2θ is the angle subtended by a grain boundary segment at the center of the grain (see Fig. 11), $K(\theta)$ is a constant factor defined in the Appendix, $r = \delta R/R$ is the normalized change in radius, and $t = h/R$ is the normalized surface layer thickness. Thus a value of $S/S_0 < 1$ would indicate a reduction in grain boundary area as a result of the morphology change, which is energetically favored. It should be noted that even though equation (2) is derived for the case of an expanding grain, it can also be applied to the case of a shrinking grain by letting δR, or r, have a negative value. This function is plotted in Fig. 9 for the cases of grain boundary segments of $2\theta = 45°$ and $120°$ respectively. Notice that these two cases correspond to, respectively, a grain with eight equal sides that is known to grow, and one with three equal sides that is known to shrink.

108

(a) 8 sided grain   (b) 3 sided grain

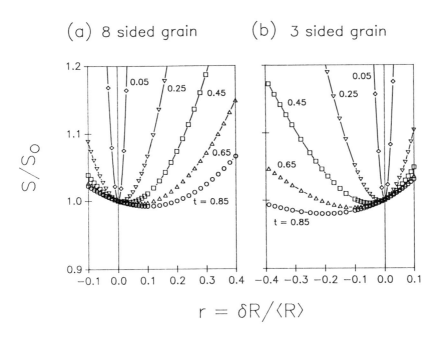

$$r = \delta R / \langle R \rangle$$

Fig. 9   Plots of S/S_0 versus r, where $r = \delta R/R$. The curves are calculated
from equation (2) with various t, where $t = h/R$. The angles subtended
by the grain boundary segment are $2\theta =$ (a) $45°$, (b) $120°$, which
correspond to grains with eight and three equal sides respectively.

Let us first consider the case of the eight sided grain shown in Fig.
9(a). Since h is a constant, the smaller R value typical of the fine grains
in the initial stage of film growth results in a larger t value, say 0.85.
The system is thus on the t = 0.85 curve initially. Moving to positive r
($\delta R > 0$, which is equivalent to the grain expanding) from the origin
results in a $S/S_0 < 1$. In other words, a growing grain results in less
grain boundary area. This provides the driving force for the grain to grow.
As the film grows thicker and the grain radius R becomes larger, t becomes
smaller. The system then moves up the family of curves gradually towards
those with smaller t values, until R reaches a critical value which give a
very small value of t, say, 0.05. This curve has its minimum very close to
r = 0 such that any change in r results in a S/S_0 value larger than one.
The grain boundary motion then ceases at this point. The exact opposite
situation is obtained by examining the case of the three sided grain shown
in Fig. 9(b), where the grain would shrink at the initial stage of film
growth causing the system to move down the family of curves towards larger
t values, which then results in further shrinking of the grain.

The slopes of the curves in Fig. 9 give the rate at which S/S_0 changes
with r, and are thus a direct measure of the magnitude of the driving force
for the grain expansion. The driving force is zero at the minima of the
curves where the slopes are zero. These minima are obtained by setting the
first derivative of equation 2 to zero, which results in the following
relationship

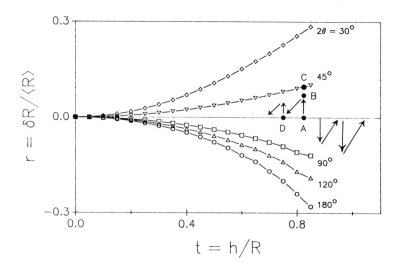

Fig. 10 The locus of the minima of equation (2) plotted as r versus t for
various 2θ angles. The saw-tooth shaped path above the x-axis show
schematically the path taken by an expanding $2\theta = 45^\circ$ grain boundary
segment, and the other below the x-axis show that for a shrinking 2θ
$= 120^\circ$ segment.

$$2r^2 + 2r + t^2 - t \ (r^2 + t^2)^{\frac{1}{2}} / K(\theta) = 0 \qquad\qquad (3)$$

Fig. 10 shows this function as a plot of r versus t for grain boundary segments with various 2θ. This figure gives us a different perspective on the
time evolution of the thin film system. Let us consider the case of $2\theta =$
45°, which corresponds to the case of an eight equal sided grain. At the
beginning of a fresh Monte Carlo Step (MCS) the system is represented by
the point labled 'A'. This point is on the x-axis, where $r = \delta R/R = 0$, as a
result of the fact that $\delta R = 0$ at the begining of a new MCS. The driving
force would then push the system to 'B' in the direction of the minimum
'C'. The value of R is thus increased to $R' = R + \delta R$. This then yields a
new and smaller $t = h/R'$, which puts the system at 'D' at the beginning of
the next MCS. The evolution of this system would then follow a saw-tooth
shaped path shown in Fig. 10. Similar arguments for the case of $2\theta = 120^\circ$
show that this system would shrink away following a different saw-tooth
shaped path, which is below the x-axis, until the grain completely van-
ishes. The criterion for a grain growing or shrinking can thus be established from Fig. 10 as follow: a grain grows if its curve is above the
x-axis, and shrinks if below. In other words a grain grows if it has more
than six equal sides, and shrinks if it has fewer than six equal sides.

We can also establish the grain size at which the grain boundary move-ment of the larger grains are pinned. This takes place when the curves reach the x-axis, where $\delta R = 0$, in Fig. 10. In the frame work of our pre-sent study, however, the microstructure is mapped onto is a discrete lat-tice, such that δR cannot approach zero indefinitely. Instead, there is a lower bound of the value that δR can take, which is the lattice spacing, a_0. Substituting a_0 for δR in equation (3), we get the critical cross-sectional radius, R_1, at which the grains boundary movement of an expanding grain ceases

$$R_1 = (h^2/2a_0) \{ \sqrt{(1+[a_0/h]^2)}/K(\theta) - 1 \} - a_0 \tag{4}$$

Similarly, using $\delta R = -a_0$ we can get R_2 at which the shrinking grain is pinned

$$R_2 = (h^2/2a_0) \{ \sqrt{(1+[a_0/h]^2)}/K(\theta) + 1 \} + a_0 \tag{5}$$

Notice that equation (5) determines the smallest radius a large grain can shrink down to.

In this work, the thickness of the top layer in which the microstruc-ture is allowed to evolve is equal to three lattice spacing, $h = 3a_0$, which is the interaction range (see equation (1)) used in the simulation. The critical radii can then be calculated by substituting this value into either equation (4) or (5), and are found to be 0.68 a_0 and 3.56 a_0 for an expanding grain with eight and twelve equal sides respectively. These val-ues are of the same order of magnitude as the actual $<R> = 4a_0$ observed.

A closer examination of the pinned microstructure shown in the cross-sectional top view in Fig. 2 reveals that all the large grains have more than six sides and no small grains have three or fewer sides, which is con-sistent with our analysis. There are, however, many small grains with six or fewer sides that have not yet shrunk away. The stabilization of these small grains may be a consequence of topological transformations. These grains start out as expanding grains with more than six sides. As they grow larger, however, they loss a number of sides when some of their shrinking neighbors vanish. By this time, nevertheless, these grains have acquired radii which are larger than the limiting minimum values dictated by equa-tion (5). As a result they can no longer shrink down.

It would be interesting to see if we can obtain the grain size of a real columnar structured thin film from equation (4). Similar to this simu-lation, a real thin film is not a continuum but has a discrete limit, which is its atomic spacing. The thickness of the surface layer in which the grain boundaries can move depends on the deposition conditions. If we assume this thickness to be 200 atomic layers, with a typical atomic spac-ing of 4Å, the critical radius for an eight sided grain can then be calcu-lated from equation (4). This has been done and is found to be approxi-mately 2.4 μm, which is a very reasonable value [18,21].

Summary

1. A computer simulation procedure based on the Monte Carlo algorithm has been developed to model thin film microstructure evolution in the Zone II regime.

2. The cross-sectional grain size approaches a limiting value with time. Once a critical size has been reached, the grain boundaries become pinned, resulting in the formation of a columnar morphology for the thin film.

3. The cross-sectional grain size distribution in the pinned state is approximately log-normal.

4. An analytical model is proposed in which the pinned state is accounted for in terms of a grain boundary drag at the growing interfaces due to sub-surface boundaries.

5. The introduction of anisotropic solid-vapor interfacial energy can unpin the grain boundaries, and induce abnormal grain growth.

6. The grain boundary movement again ceases and forms a new columnar grain structure when the abnormal grains impinge upon one another.

Appendix:

In this Appendix we would like to calculate the grain boundary area of the grain morphologies shown in Fig. 8 using the Feltham construction [20]. Fig. 11 is a schematic diagram of a grain boundary segment which is subtending an angle of 2θ at the center of the grain. It is anchored at both ends and is assumed to intersect other grain boundaries at these ends at an equilibrium 120° angle. The radius of curvature, ρ, of the grain boundary segment is

$$\rho = 2 R \sin \theta / (\sqrt{3} \sin \theta - \cos \theta) \qquad (6)$$

where R is the radius of the grain. The length of the grain boundary is then

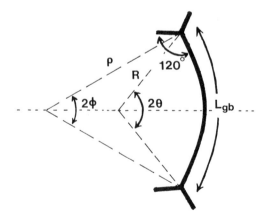

Fig. 11 Schematic diagram of a grain boundary with length L_{gb} anchored at both ends, where it intersects other grain boundaries at 120° angles. The angle subtend by the grain boundary at the grain center and the center of curvature are 2θ and 2ϕ, respectively. The grain radius and the radius of curvature of the boundary are R and ρ respectively.

112

$$L_{gb} = 2\phi \ \rho$$

$$= 4 \ R \ (\theta - 60°) \ Sin \ \theta \ / \ (\sqrt{3} \ Sin \ \theta - Cos \ \theta)$$

$$= K(\theta) \ R \qquad\qquad (7)$$

where L_{gb} is the length of the grain boundary, 2ϕ the angle the grain boundary subtends at the center of curvature, $K(\theta)$ a constant factor whose value depends on θ only. The grain boundary area of the original morphology, S_o, is thus found to be

$$S_o = K(\theta) \ R \ h \qquad\qquad (8)$$

To obtain the grain boundary area of the new morphology, S, we have to take into account both the now tilted grain surface as well as the 'flaps' (see Fig. 8) consumed by the growing grain. Since there are as many flaps attached to a grain as the number of sides a grain has, each side on the average gets to consume one flap. Assuming a grain radius that decreases linearly with height, from R at the bottom of the layer to R + δR at the top, the surface area of the grain is obtained by integrating equation (7) from the bottom of the layer to the top. Taking into account the triangular shaped flaps that have been consumed, the grain boundary area of the new morphology, S_o, is found to be

$$S = K(\theta) \ (2R + \delta R) \ \sqrt{(h^2 + \delta R^2)} \ / \ 2 \quad - \quad h \ \delta R \ / \ 2 \qquad\qquad (9)$$

where the first term is from the tilted grain boundary surface, and the second term from the flap.

References:

1. B.A.Movchan and A.V.Demschishin, Phys.Met.Metallogr. <u>28</u>, 83 (1969).

2. J.A.Thornton, Ann.Rev.Mater.Sci. <u>7</u>, 239 (1977).

3. R.Messier, A.P.Giri, and R.A.Roy, J.Vac.Sci.Technol. <u>A2</u>, 500 (1984).

4. P.Meakin, P.Ramanlal, L.M.Sander, and R.C.Ball, Phys.Rev. A <u>34</u>, p.5091 (1986)

5. D.J.Srolovitz, to be published.

6. D.Weaire and J.P.Kermode, Phil.Mag.B <u>47</u>, L29 (1983); <u>48</u>, 245 (1983); <u>50</u>, 379 (1984).

7. J.Wejchert, D.Weaire and J.P.Kermode, Phil.Mag.B <u>53</u>, 15 (1986).

8. H.J.Frost, C.V.Thompson, C.L.Howe and J.Whang, to be published in Scripta Metallurgica (1988).

9. M.P.Anderson, D.J.Srolovitz, G.S.Grest and P.S.Sahni, Acta Met. <u>32</u>, 783 (1984).

10. M.P.Anderson, D.J.Srolovitz, G.S.Grest, and P.S.Sahni, Acta Met. <u>32</u>, 783 (1984)

11. D.J.Srolovitz, M.P.Anderson, P.S.Sahni, and G.S.Grest, Acta Met. <u>32</u>, 793 (1984).

12. D.J.Srolovitz, J.Vac.Sci.Technol.A <u>4</u> (6), 2925 (1986).

13. F.Haessner and S.Hofmann, in <u>Recrystallization of Metallic Materials</u>, ed. F.Haessner, Dr. Riederer Verlag GmbH, Stuttgart, 76 (1978).

14. J.L.Walter and C.G.dunn, Acta metall. <u>8</u>, 497 (1960)

15. F.H.Buttner, E.R.Funk and H.Udin, J. Phys. Chem. <u>56</u>, 657 (1952).

16. D.Kohler, J. Appl. Phys. suppl. <u>31</u>, 408S (1960)

17. D.S.Rickerby, G.Eckold, K.T.Scott, and I.M.Buckley-Golder, Thin Solid Films <u>154</u>, 125 (1987).

18. A.F.Jankowski, Thin Solid Films <u>154</u>, 183 (1987).

19. C.V.Thompson, presentation given at TMS annual meeting, Phoenix, 1988. To be published in conference proceedings.

20. P.Feltham, Acta Met., <u>5</u>, 97 (1957).

21. M.C.Madden, Thin Solid Films <u>154</u>, 43 (1987).

OBSERVATIONS OF GRAIN GROWTH IN THIN FILMS

C.V. Thompson, Dept. of Materials Science and Engineering,

M.I.T., Cambridge, MA   02139

Abstract

Recent experimental characterization of grain growth in metallic and semiconductor thin films is reviewed. In films with initial grain sizes less than the film thickness, normal grain growth leads to columnar structures in which the grain boundary planes are roughly perpendicular to the plane of the film and the average in-plane grain diameter is 2-4 times the film thickness. Grain size distributions in these structures are usually log-normal. Subsequent grain growth typically requires higher annealing temperatures and occurs through abnormal or secondary grain growth in which a fraction of the grains grow into a matrix of static normal grains. Secondary grain growth in thin films often leads to films with restricted or uniform texture. Secondary grain growth can occur during deposition of metallic films at temperatures as low as 20% of their absolute melting temperatures. Impurities can promote secondary grain growth through precipitate formation. Other factors affecting the kinetics of grain growth are briefly discussed.

Microstructural Science for Thin Film
Metallizations in Electronics Applications
Edited by J. Sanchez, D.A. Smith and N. DeLanerolle
The Minerals, Metals & Materials Society, 1988

Introduction

In a wide variety of electronic applications, the electronic properties and stability of polycrystalline films are strongly affected by their grain size, grain orientations and grain size distributions. These are affected by grain growth which occurs during deposition of the film or during post-deposition processing. Grain growth in thin films has several important distinctive features compared to grain growth in bulk materials. Over the past few years, a number of experimental studies of grain growth in thin films have been carried out at M.I.T. In this paper, these studies will be briefly reviewed and summarized in order to present a general overview of the characterisitics of grain growth in thin films and in order to also indicate some of the ways in which grain growth in thin films can be controlled.

The initial microstructure of a polycrystalline film can vary widely, depending on the material type, the deposition technique and the deposition conditions. Grains often have average in-plane dimensions, d_i, which are significantly smaller than the average dimensions perpendicular to the plane of the film, d_p (Fig. 1a). This is true, for example, of chemical vapor deposited crystalline silicon films[1] and metallic films deposited at low substrate temperatures[2]. However, crystallized amorphous silicon and metal films deposited at higher temperatures generally have roughly equiaxed grains. If the as-deposited grains of a film are equiaxed and smaller than the film thickness, grain growth is initially 3-dimensional and should have characteristics identical to bulk normal grain growth.

Normal Grain Growth in Germanium Films

We have characterized 3-D normal grain growth in germanium films deposited by electron-beam evaporation onto amorphous SiO_2 layers which were thermally grown on silicon substrates[3]. When deposited at room temperature, the germanium films were amorphous. Films were crystallized and grain sizes were determined using dark field plan view transmission electron microscopy after annealing for various times at various temperatures. Figure 2 shows the measured average grain size as a function of annealing time at 775°C for a 500 Å-thick film. Grain growth has clearly occurred. However, given the relationship

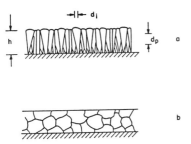

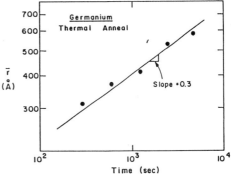

Figure 1. Schematic drawings of cross sections of films with a) columnar grains and b) equiaxed grains.

Figure 2. The log of the normal grain size versus the log of the annealing time at 775°C. Results are for a 500Å-thick pure germanium film on SiO_2.[3]

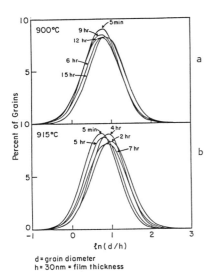

d = grain diameter
h = 30nm = film thickness

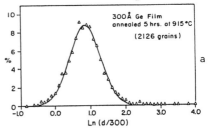

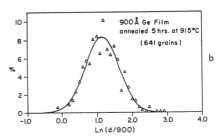

Figure 3. Grain size distributions for a) 300Å-thick and b) 900Å thick germanium films annealed for 5 hours at 915°C.[6]

Figure 4. Grain size distributions for 300Å-thick germanium films annealed for various times at a)900°C and b) 915°C.[6]

$$d^m - d_o{}^m = \alpha t$$

where d is the average grain diameter, d_o is the initial average grain diameter, α is a temperature dependent constant and t is time, the measured value of m is approximately 3 instead of 2 as expected from theory[4]. This may be the result of changes in the growth rate that occur as d approaches the film thickness h. Similar experiments on phosphorus doped silicon have yielded values of m much closer to two.[5]

After the average grain diameters have increased to sizes comparable to the film thickness, a columnar structure develops, i.e., all grain boundaries intersect both the top and bottom surfaces of the films. At this point, normal grain growth stops. This saturated columnar structure has been characterized in detail for germanium by Palmer et al.[6] As can be seen in figure 3, the grain size distribution is log-normal and the ratio of the average grain diameter and the film thickness, d/h, is roughly 2.5, independent of film thickness. Once this structure has developed, no further normal grain growth occurs, as can be seen in figure 4, even though a driving force for 2-dimensional normal grain growth still exists. This saturation phenomenon has been observed in a variety of films and sheet and has been called the "specimen thickness effect" by Beck.[7] While 2-dimensional normal grain growth has been extensively modelled[4,8-9], there appear to be no experimental observations of this mode of grain growth in thin films beyond d $\cong$ 2.5h.

It should be noted that the log-normal grain size distributions observed in saturated normal grain structures differ significantly from distributions predicted by Hillert[4] and Louat [10] for steady state normal grain growth, as illustrated in figure 5. Log-normal distributions are also generally observed in unsaturated 3-D normal grain structures.[11-13] Recognizing this empirically determined nature of normal grain growth, models which predict log-normal or similar distributions have recently been proposed.[15-16]

Secondary Grain Growth in Semiconductor Films

When 2-dimensional grain growth does occur in thin films, it occurs through abnormal or secondary grain growth. In this mode of grain growth, a few grains grow at rates that are much higher than the average growth rate. This initially results in a bimodal grain size

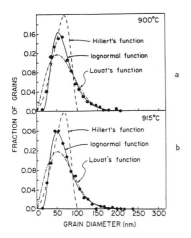

Figure 5. Normal grain size distri-
distributions measured in 300Å-thick
germanium films annealed for 5
minutes at a) 900°C and b) 915°C.
Compared to distributions predicted
by Hillert and Louat.[6]

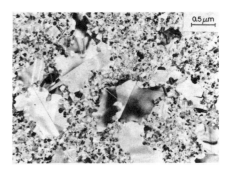

Figure 6. Transmission electron
micrograph showing normal and
secondary grains in a 300Å-thick
germanium film annealed for
5 hrs. at 915°C.[17]

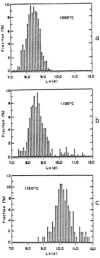

Grain Size Distributions

Figure 7. Grain size distributions
for 1140Å-thick silicon films doped
with 1×10^{21} cm^{-3} with phosphorous
and annealed for 20 min. at a)1050°C
(normal grains), b) 1100°C (normal
and secondary grains),and c) 1150°C
(secondary grains).[18]

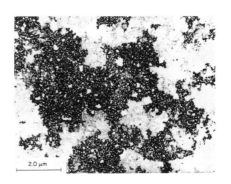

Figure 8. Typical grains in a 150Å-
thick germanium film annealed for 5
hours at 900°C. Large regions of
uniform contrast (~10μm) are single
grains. Smaller bright spots
are holes in the film.[6]

distribution as is illustrated by the electron micrograph[17] in figure 6 and by the measured grain size distributions[18] in figure 7. Eventually, secondary grains consume the entire film and a monomodal distribution of large grains develops.

Clearly, 2-dimensional normal grain growth is suppressed in thin films and secondary grain growth occurs when a few grains overcome the restraint on normal grain growth. However, it is not clear what causes stagnation of 2-D normal grain growth. It seems likely that at least in some systems, grain growth is impeded by development of grooves at grain boundary surface intercepts. This was originally proposed by Mullins[19] who also pointed out that surface energy reduction can provide a driving force for selective motion of grooved boundaries between grains with sufficiently different free surface energies. As a result, grains with orientations which lead to minimum surface or interface energies will be most likely to break free and grow as secondary grains. This is consistent with the fact that secondary grain growth in thin films generally leads to development of restricted or uniform textures, as observed for example in germanium[6], silicon[20] and gold[21].

Two-dimensional secondary grain growth is driven by the elimination of grain boundaries and by the reduction of the surface and/or interface energies of a film. It can be shown that the rate of secondary grain growth can be given by[22,23]

$$\dot{r}_s = M \left(\frac{2 \, \gamma_{gb}}{r_n} + \frac{\Delta\gamma}{h} \right)$$

where $\dot{r}_s$ is the rate of change of the average secondary grain radius, r_n is the average normal grain radius, M is the grain boundary mobility, γ_{gb} is the average grain boundary energy and $\Delta\gamma$ is the difference of the average surface energy of normal grains and the average surface energy of secondary grains. Because of the specimen thickness effect, $r_n \cong h$ so that at a given temperature, the rate of secondary grain growth should increase with decreasing film thickness. Also, in ischronal anneals the temperature required for secondary grain growth should decrease with decreasing film thickness. This has been observed to be the case in germanium[6] and silicon[18] films.

The final secondary grain size is a function not of the rate of growth but of the number of grains in the initial structure that can break free and become secondary grains. This, in turn, depends on the distribution of sizes and orientations of the initial grains as well as the detailed nature of the constraint on normal grain growth. The final grain size can be smaller in thinner films, due to the higher areal concentration of normal grains in the initial structure.[18] However, in very thin, partially discontinuous germanium films we have observed growth of very large grains (Fig. 8), presumably because fewer of the initial grains can grow past the holes in the film.

Secondary Grain Growth in Metallic Films

Grain growth in semiconductor and metallic films is phenomenologically very similar; secondary grain growth leads to large textured grains after normal grain growth stagnates. However, the temperature required for grain growth in Si and Ge is about 60% of the absolute melting temperature. In contrast, grain growth in metals can occur at 20% of the absolute melting temperature. This might be anticipated in view of the significantly lower self-diffusities observed in diamond cubic semiconductors compared to close packed metals.[24]

An example of low temperature grain growth in a metallic film is illustrated by the series of pictures in figure 9.[21] These plan view transmission electron micrographs show 3 stages of microstructural evolution in gold films which have been deposited on Si wafer coated with thermally grown amorphous SiO_2. The gold deposited using electron beam evaporation onto room temperature substrates. In figure 9a, the approximate film thickness is 100 Å and the film is not yet fully continuous. At this point, the film is composed of small grains which do not appear to be textured. As the film thickened and became more continuous, secondary grains began to grow and consume the film. These secondary grains had (111) texture. It should be noted that secondary grain growth continued after deposition and well after any heat generated during deposition should have been dissipated.[21]

Clearly in the case of Au on SiO_2, grain growth during deposition affects the initial structure of the films, even when the substrate temperature is only 20% of the absolute melting temperature. This reinforces the suggestion by Grovenor et al[25] that low temperature grain growth must be considered in any generalized picture of the evolution of microstructures in thin films. In the context of the zone model[2], it should be noted that the transition from zone I to zone II morphology occurs at about 0.1 to 0.3 T_m. This transition might be associated with

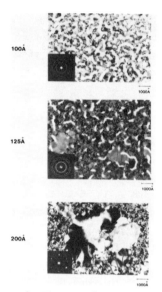

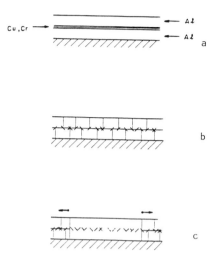

Figure 9. Electron beam evaporated
gold deposited on SiO_2 at room
temperature. Nominal film thicknesses are indicated at the left.[21]

Figure 10. Schematic drawings of
cross-sections of a) as deposited
Al, Cu, Cr, Al films and b) the
microstructure after annealing to
form precipitates. c) Precipitates
impede normal grain growth until
secondary grain growth occurs.

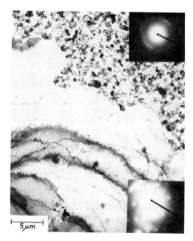

Figure 11. Secondary grain growing
in a Al - 2% Cu - 0.3% Cr film.[27]

grain growth during the early stages of deposition. The driving force
for secondary grain growth is very high during coalesence of a
continuous film. This is due to both the small initial grain size and
the high surface-to volume ratio. During subsequent deposition,
epitaxial growth on stagnant secondary grains can lead to columnar
structures.

Use of Precipitates to Promote Secondary Grain
Growth in Thin Films

Some years ago, Gangulee and d'Heurle[26] demonstrated that addition
of alloy elements to aluminum can promote growth of very large secondary
grains. We have recently reproduced their results for Al - 4% Cu - 0.3%
Cr films.[27] These films were formed through sequential deposition of
the elemental components. Cu and Cr films were deposited between pure
Al films of equal thickness, as illustrated in Fig. 10a. When deposited
at room temperature, the top Al films have strong (111) textures.
Annealing of the composite films leads to formation of precipitates
which block normal grain growth. These precipitates form mostly in the
center of the film as illustrated in Fig. 10b. At temperatures between
400 and 450°C, a few grains break free of the precipitates (see Fig. 10c
and Fig. 11). The resulting secondary grains were found to have (112)
or (110) texture.

In the case described above, normal grain growth is impeded by
precipitates instead of grain boundary grooves. However, secondary
grains still had preferred texture even though the films were relatively
thick (0.75μm) and even though it is likely that their initial texture,
(111), already minimized the surface and interface energy of the film.
It is possible in this case that secondary grains have specific texture
because the normal grain matrix is textured and the secondary grain
boundaries are special boundaries with high mobilities.

Enhancement of Grain Growth in Thin Films

Our work on dopant-enhanced and ion-bombardment-enhanced grain
growth has recently been reviewed and correlated.[28] Here the results
will be only briefly summarized.

It has been known for some time that n-type dopands lead to
increased grain growth rates in silicon [29-31] while p-type dopants have

123

little or no effect.[30,31] We have recently carried out a detailed series of experiments on the effect of dopants on 3-D normal grain growth [5,32] and secondary grain growth in silicon.[33,18] The vacancy concentration in silicon is a known function of the free electron concentration[34] and we have shown that increased grain growth rates accompanying doping of silicon can be quantitatively accounted for if it is assumed that grain boundary mobilities are a function of the vacancy concentration.[18,5,32]

We have also demonstrated that ion bombardment of thin films can lead to grain growth at temperatures well below those required without bombardment.[3,35] The activation energy for ion bombardment enhanced grain growth (IBEGG) in silicon is about 0.1eV and the rate of IBEGG scales directly with the rate of point defect generation in Ge, Si and Au.[35] These and other observations have lead us to propose that IBEGG occurs due to extrinsic point defect generation during ion bombardment.[36]

In general, it seems likely that processes or conditions that lead to point defect generation in thin films should also lead to enhanced grain growth.

Summary

Recent work on grain growth in thin films has been reviewed. This review has focused primarily on work carried out at M.I.T. From this work and the work of others, several general conclusions can be tentatively drawn:

1) In thin films, 3-D normal grain growth generally leads to formation of a stagnant 2-D columnar structure in which grain sizes are log-normally distributed and the average grain radius is roughly equal to the film thickness.

2) Subsequent secondary grain growth leads to larger grains which usually have restricted textures.

3) In the absence of texture in the normal grain matrix, texture development among secondary grains likely results from surface energy minimization.

4) Precipitates can impede normal grain growth in films, and also lead to secondary grain growth.

5) In the early stages of deposition of metallic films, grain growth can affect microstructural evolution at as low as 20% of the absolute melting temperature of a film.

6) Processes or conditions which lead to increased point defect concentrations, e.g. dopants in semiconductors or ion bombardment in general, can lead to enhanced grain growth rates.

References

1) E. Kinsbron, M. Sternheim, and R. Knoell, Appl. Phys. Letts. 42, 835 (1983).

2) J.A. Thorton, Ann. Rev. Material Sci. 7,239 (1977).

3) H.A. Atwater, H.I. Smith and C.V. Thompson, Materials Research Society Symposium Proceedings 31,337 (1985).

4) M. Hillert, Acta Metall. 13, 227 (1965).

5) H.J.- Kim and C.V. Thompson, submitted to J. Electrochem Soc.

6) J. Palmer, C.V. Thompson, and H.J. Smith, J. Appl. Phys. 62, 2492 (1987).

7) P.A. Beck, M.L. Holtzworth, and P.R. Sperry, Trans. AIME 180, 163 (1949).

8) M.P. Anderson, D.J. Srolovitz, G.S. Grest, and P.S. Sahni, Acta Metall. 32, 783 (1984).

9) H.J. Frost, C.V. Thompson, C.L. Howe, J. Whang, Scripta Metall. 22, 65 (1988).

10) N.P. Louat, Acta Metall. 22, 712 (1974).

11) P.A. Beck, Philos. Mag. Suppl. 3, 245 (1954).

13) D.A. Abouv and T.G. Langdon, Metallography $\underline{2}$ 171 (1969).

14) H. Conrad, M. Swintowski and S.L. Mannan, Metall. Trans. $\underline{16A}$.

15) S.K. Kurtz and F.M.A. Carpay, J. Appl. Phys. $\underline{51}$, 5725 (1980)

16) C.S. Pande, Acta Metall. $\underline{35}$, 2671 (1987).

18) H.J. Kim, Ph.D. Thesis, Department of Materials Science and Engineering, Massachusetts Institute of Technology (1988).

19) W.W. Mullins, Acta Metall. $\underline{6}$, 414 (1958).

20) C.V. Thompson and H.I. Smith, Appl. Phys. Letts. $\underline{44}$, 603 (1984).

21) C.C. Wong, H.I. Smith, and C.V. Thompson, Appl. Phys. Letts. $\underline{48}$, 335 (1986).

22) C.V. Thompson, J.Appl. Phys. $\underline{58}$, 763 (1985).

23) C.V. Thompson, to appear in Acta Metall.

24) A.M. Brown and M.F. Ashby, Acta Metall. $\underline{28}$, 1085 (1980).

25) C.R.M. Grovenor, H.T.G. Hentzwell and D.A. Smith, Acta Metall. $\underline{32}$, 773 (1984).

26) A. Gangulee and F. d'Heurle, Thin Solid Films $\underline{16}$, 227 (1973.

27) Thompson, C.V. and Maiorino, C.D., Proceedings of the 1986 Int'l Conf. on Solid State Devices and Materials, Tokyo, Japan, p. 491 (1986).

28) C.V. Thompson, in "Diffusion Processes in High Technology Materials," ed. by D. Gupta and A.D. Romig, Jr., Symposium of TMS/AIME fall meetings, Cincinnati, OH, 1987.

29) Y. Wada and S. Nishimatsu, J. Electrochem. Soc. $\underline{125}$, p. 1499 (1978).

30) L. Mei, M. River, Y. Kwark and R.W. Dutton, J. Electrochem. Soc. $\underline{129}$, 179 (1982).

31) R. Angelucci, M. Severi, and S. Solmi, Mater. Chem. Phys. $\underline{9}$, 235 (1983).

32) H.J. Kim, and C.V. Thompson, proceedings of the Fall 1987 Materials Research Society Symposium on Polysilicon Films and Interfaces, Boston, MA.

33) H.J. Kim and C.V. Thompson, Appl. Phys. Letts. $\underline{48}$, 399
    (1986).

34) J.A. Van Vechten and C.D. Thurmond, Phys. Rev. $\underline{B14}$, 3539
    (1976).

35) H.A. Atwater, C.V. Thompson and H.I. Smith, Materials
    Research Society Symposium Proceedings $\underline{74}$, 499 (1987).

36) H.A. Atwater, C.V. Thompson, and H.J. Smith, Phys. Rev.
    Letts. $\underline{60}$, 112 (1988).

INTERACTION OF POINT DEFECTS WITH GRAIN BOUNDARIES

Marek Dollar

Metallurgical Engineering and Materials Science Department
Carnegie Mellon University
Pittsburgh, PA 15213

permanent address:
Department of Metallurgy
Academy of Mining and Metallurgy
Krakow, Poland

Abstract

The present state of understanding of the interaction of grain boundaries with point defects is summarized. First, selected experimental results are reviewed in order to illustrate the significance of the interaction for a number of processes in polycrystals. Then, dislocation and atom models for grain boundaries as point defect sinks/sources are described. Finally, attention is focussed on the influence of the atomic structure of grain boundaries on their efficiency as sinks/sources for point defects in the light of the recent experiments. It has been shown that random high angle boundaries are more efficient than special high angle as well as low angle grain boundaries. Experimental evidence is also presented which shows that this interaction affects mechanical properties.

Microstructural Science for Thin Film
Metallizations in Electronics Applications
Edited by J. Sanchez, D.A. Smith and N. DeLanerolle
The Minerals, Metals & Materials Society, 1988

Introduction

There is considerable experimental evidence indicating that grain boundaries in bulk materials act as point defect sinks or sources. The generation and annihilation of point defects is of significance for a number of processes in polycrystals, such as diffusional creep, formation of precipitate free zones, shrinkage and growth of voids during creep or irradiation, etc. (for review, see [1,2]). Grain boundary sink/source action is often coupled with grain boundary migration.

In the present paper we first consider some aspects of the sink/source behavior of (stationary) grain boundaries. Attention is focussed on a relationship between the structure of grain boundaries and their efficiency as point defect sinks/sources. Theoretical arguments (3,4) suggest that the efficiency may depend on the boundary structure. However, the experimental situation seems to be controversial (5-7). Thus, we review the results of an experiment (8) which unequivocally indicate that the boundaries of low energy are less efficient point defect sinks/sources than high energy boundaries, and we try to explain this observation.

Further, we consider why the interaction of point defects with grain boundaries is of importance in thin films. Once again, we focus our attention on the role of grain boundary structure, now on its ability to reduce diffusivity in low energy boundaries. Assuming that the physical reason for both the lower point defect sink/source efficiency and the lower diffusivity of low energy boundaries is the same, we believe that the discussion to follow allows us to substantiate why textured films exhibit slower electromigration flux than randomly oriented films, a phenomenon well known in thin film devices (9).

Grain Boundaries As Point Defect Sinks/Sources

The interaction of grain boundaries with point defects can be exemplified by Figure 1 showing a dislocation loop depleted zone near a grain boundary in quenched aluminum. The dislocation loops are created by quenching. The annihilation of vacancies in grain boundaries leads to the formation of zones denuded of loops.

The easiest way to visualize the sink (source) behavior of a grain boundary is to assume that point defects are annihilated (created) by the climb of dislocations in the boundary. In fact, several transmission electron microscopy studies have proven that there is a correlation between the climb of dislocations occurring in a boundary and a flux of point defects from or to the boundary (10, 11). In consequence, many models for grain boundaries as point defect sinks/sources which have been proposed (for review see [1,2]) are based on this assumption.

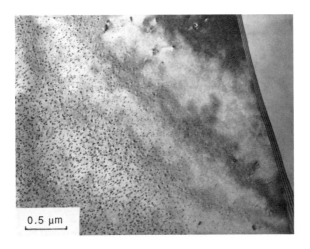

Figure 1 - Electron micrograph of quenched and aged aluminum
showing a dislocation loop-depleted zone near a grain boundary.

A detailed discussion on grain boundary structure is not the focus of this paper.
However, in order to facilitate further discussion, it seems essential to recall the idea of
modelling of grain boundaries in terms of arrays of closely spaced dislocations. Taking
this microstructural aspect into account, we can then classify grain boundaries (see Table
I), differentiating between low angle (where the angular misorientation θ across the
boundary is < 10°) and high angle boundaries (θ > 10 - 15 °). The high angle boundaries
may be divided into special (either exact or near coincidence) and random boundaries. This
classification of high angle boundaries reflects the concept of coincidence sites in grain
boundaries (for review see [12,13]), i. e. sites at which the lattices of the two crystals
forming a boundary would coincide if they were extended to the other side of the
boundary. The density of the coincidence-site lattice depends on the orientation
relationship between the crystals. The boundaries between the crystals forming the
coincidence-site lattice with a high density of coincidence sites are called special
coincidence boundaries. The low angle and special high angle boundaries are of relatively
low energy as compared to random ones.

Except for the random cases, grain boundaries can be represented as arrays of grain
boundary dislocations (GBDs). Low angle and exact coincidence boundaries may be
described as containing primary (lattice) grain boundary dislocations (PGBDs). In near
coincidence boundaries secondary grain boundary dislocations (SGBDs) are required to
accomodate deviations from exact coincidence misorientations. It is presently generally
accepted that the above mentioned boundaries act as point defect sinks/sources by the climb
of GBDs, either primary or secondary.

Table I. The Classification Of Grain Boundaries

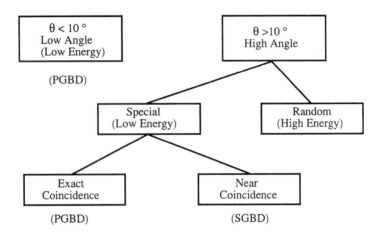

Formally, GBD models can be extended to include random boundaries. Any random boundary can be thus represented as an array of very closely spaced dislocations. Thus, it has been proposed that all grain boundaries act as point defect sources/sinks by the climb of GBDs (2). Such an approach seems to be justified only if all boundaries are effective sinks/sources. In fact, the results of many experiments suggest that all boundaries (except coherent twin interfaces) are equally effective sinks/sources (6,7). However, the experimental situation is controversial, since the results of other experiments suggest an influence of grain boundary structure (energy) on the efficiency of grain boundaries as point defect sources/sinks (5,8). Below we report such an observation.

The material used was 99.99% polycrystalline gold. The specimens (50 µm thick foils) were annealed to stabilize the grain structure. Then, the foils were bombarded with argon ions. The bombardment had a twofold purpose: (i) to generate a large concentration of point defects, and (ii) to reduce the specimen thickness to values suitable for transmission electron microscpy (TEM). TEM in situ isothermal annealing studies were then performed.

The typical structure of as-irradiated gold is shown in Figure 2. A large and uniform density of dislocation loops, resulting from the condensation of point defects, were observed both inside grains and in the vicinity of grain boundaries. During annealing, two types of boundaries were noticed. In the vicinity of certain boundaries loop shrinkage occurred during in situ annealing in the temperature range 180 °C - 220 °C, as shown in Figure 3. The boundaries were found to be random high angle boundaries. The second group consists of boundaries in the vicinity of which preferential loop shrinkage did not occur at any temperature (above 220 °C loop shrinkage was observed near the boundaries as well as in the grain interior, as illustrated in Figure 4).

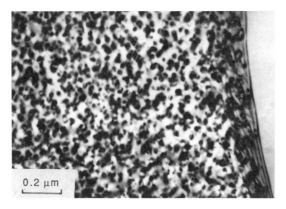

Figure 2 - Electron micrograph showing dislocation loops in as-irradiated gold foils.

Figure 3 - Same area of the specimen shown in Figure 2 after in situ annealing for 5 minutes.

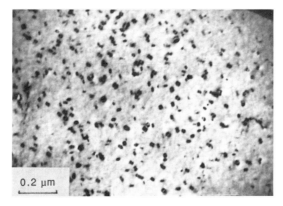

Figure 4 - Electron micrograph of an irradiated gold foil annealed for 1 minute at 220 °C. The boundary in the upper right corner corresponds to the boundary #8 in Table II.

Table II. Boundaries near which loops were not seen to disappear during in situ annealing of thin foils.

Boundary Number	Misorientation Angle (°)	Axis Of Rotation
1	2	100
2	3	100
3	4	100
4	4	100
5	4	110
6	4	110
7	7	110
8	16 (Σ= 25)	100
9	22 (Σ= 13)	100
10	11 (Σ= 41)	100
11	13 (Σ= 73)	110
12	14 (Σ= 73)	110
13	16 (Σ= 51)	110

Table II lists the boundaries near which loops were not seen to disappear, characterized by misorientation angle and rotation axis. As indicated below, these are either low angle or special high angle boundaries. The boundaries 1-7 are low angle grain boundaries; boundaries 8 and 9 are special, high angle boundaries with the reciprocal coincidence site density Σ = 25 and 13, respectively (the quantity Σ is defined as the reciprocal of the fraction of lattice atoms associated with coincidence-site lattice points). Boundaries 10, 11, 12, and 13 can be formally treated as possessing coincidence lattices with Σ = 41, 73, 73, and 51, respectively. A question arises about treating boundaries 10-13, which exhibit a very low degree of coincidence, as special ones. A fortuitous, if unintentional, answer to this question was provided by Ichinose and Ishida (14). These authors performed high resolution electron microscopy of grain boundaries in gold and observed that two tilt boundaries, with <110> rotation axes and rotation angles of 16 ° and 21° respectively, consisted of clearly distinguishable arrays of dislocations. The former boundary is identical to the boundary 13 listed in Table II.

Thus, all boundaries listed in Table II can be represented as arrays of GBDs. Under the conditions used in the present experiment these boundaries were found not to be effective point defect sinks. This may be understood if it is assumed that an energy barrier exists for the absorption of point defects in low angle and special high angle boundaries. The physical reason for the existence of such an energy barrier can be visualized as follows. If a boundary contains an array of GBDs, the absorption of vacancies (an example of sink/source behavior) results in the climb of GBDs which occurs by the formation and motion of jogs. This process locally decreases the dislocation spacing and, thus, increases the boundary energy so that an energy barrier results for vacancy absorption.

In contrast, random high angle boundaries were found to be effective point defect sinks.

The experimentally revealed difference allows us to suggest that the two groups of boundaries are structurally different and that the formal GBD models of random boundaries as point defect sinks are not appropriate. Physically, this occurs because the Burgers vectors of GBDs become exceedingly small and the cores of these dislocations may overlap. In consequence, a random boundary energy depends little on local variations of the boundary structure.

The higher efficiency of high energy random boundaries (as revealed also in a diffusional creep experiment [5]) may be understood in terms of models attempting to determine the atomic structure of point defects in grain boundaries. As calculated by means of computer simulation (15), point defects can be absorbed in random boundaries by defect delocalization involving the generation of an avalance of atomic motions in the boundary area without significant change in the boundary energy. No such delocalization occurs in special low energy boundaries since it would destroy the good fit. The incorporation of point defects into special boundaries results in localized defects which locally modify the energy and atomic structure of a boundary.

The Importance Of Point Defect/Grain Boundary Interaction In Thin Films

It is well known that diffusion is more rapid along grain boundaries than in the interiors of crystals. Grain boundary diffusion proceeds with a lower activation energy that lattice diffusion, therefore it becomes especially important at low temperatures. Since thin films are used at low temperatures and very often have a grain size as small as 1 nm, grain boundary diffusion should become the dominant transport mechanism in thin films. Moreover, the diffusion rate in current carrying thin films is enhanced by large current densities. In fact, it is well established that the increase of the grain size leads to the improvement of the lifetime of thin film conductors with respect to electromigration-induced failures. The dominant role of grain boundaries in thin film electromigration is nowadays generally recognized (9,16).

A question which has arisen in the past is whether a grain boundary acts as a sink for vacancies produced during lattice electromigration or whether electromigration itself is contained in the grain boundary (9). The measurements of the activation energy for the process proved that electromigration of the atoms accurs along grain boundaries (9). Lifetime studies showed also that the activation energy is higher in textured films than in random films, and that the textured films had considerably longer life (17, 18).

In analyzing grain boundary difusion it is customary to consider a grain boundary as a homogeneous slab of width w_b possessing a uniform diffusivity D_b (19). The mass transport along the grain boundary is then described by using the product parameter $w_b \, D_b$. Such an approach, though otherwise fruitful, does not allow us to discuss the difference between textured and random films.

The differences may be understood in terms of the discussed models attempting to determine the atomic structure of point defects in grain boundaries. As already mentioned, the incorporation of point defects into low energy boundaries is more difficlut than into high energy boundaries. Thus, the lower diffusivity and consequently the lower rate of electrotransport is expected in low energy boundaries. This is probably why textured films, which may be characterized by the presence of many low energy boundaries, exhibit slower electromigration flux. Thus, it is suggested here that the physical reason for both the lower point defect sink/source efficiency and the lower diffusivity of low energy boundaries is the same.

Conclusions

In irradiated gold, low angle boundaries and high angle boundaries of low energy were found to be inefficient point defect sinks whereas high energy boundaries were found to readily incorporate point defects.

These results can be understood in terms of an energy barrier associated with the absorption of point defects in low energy boundaries. No such energy barrier exists for high energy boundaries.

These results are of importance in thin films and allows us to substantiate why textured films exhibit slower electromigration flux than random films.

Acknowledgements

I wish to thank Professor H. Gleiter for fruitful discussions. This work was supported by the Alexander von Humboldt Foundation.

References

1. H. Gleiter, _Progress In Materials Science_, Chalmers Aniversary Volume, J. W. Christian et al, eds., (Oxford, U K: Pergamon Press, 1981), 125.

2. R. W. Baluffi, _Grain Boundary Structure and Kinetics_, Proceedings of American Society For Metals Science Seminar, (Metals Park, OH: ASM Publications, 1980), 297.

3. M. F. Ashby, _Scripta Metall._, 3, (1969), 837.

4. J. P. Hirth, _Metall. Trans._, 3, (1972), 3047.

5. W. Jaeger, and H. Gleiter, _Scripta Metall._, 12, (1978), 675.

6. B. K. Basu, and C. Elbaum, _Acta Metall._, 13, (1965), 1117.

7. R. W. Siegel, S. M. Chang, and R. W. Baluffi, _Acta Metall._, 28, (1980), 249.

8. M. Dollar, and H. Gleiter, _Scripta Metall._, 19, (1985), 481.

9. F. M. d'Huerle, and R. Rosenberg, _Physics of Thin Films_, vol. 7., G. Hass et al, eds., (New York: Academic Press, 1973), 257.

10. A. H. King, and D. A. Smith, _Grain Boundary Structure and Kinetics_, Proceedings of American Society For Metals Science Seminar, (Metals Park, OH: ASM Publications, 1980), 331.

11. Y. Komen, P. Petroff, and R. W. Baluffi, _Phil Mag._, 26, (1972), 239.

12. R. W. Baluffi, _Interfacial Segregation_, W. C. Johnson, J. M. Blakely, eds., (Metals Park, OH: ASM Publications, 1979), 193.

13. H. Gleiter, _Mat. Sci and Eng._, 52, (1982), 91.

14. H. Ichinose, and Y. Ishida, _J. de Physique_, Suppl. 4, C-4:39, (1985), 46.

15. W. Hahn, and H. Gleiter, _Acta Metall._, 29, (1981), 601.

16. F. M. d'Heurle, and P. S. Ho, _Thin Films- Interdiffusion and Reactions_, J. M. Poate et al, eds., (New York: John Wiley & Sons, 1978), 243.

17. M. J. Attardo, and R. Rosenberg, _J. Appl. Phys._, 17, (1970), 2381.

18. J. K. Howard, and R. F. Ross, (IBM Tech. Report 22, 1968, 601).

19. R. W. Baluffi, and J. M. Blakely, _Thin Solid Films_, 25, (1975), 363.

MICROSTRUCTURE DEVELOPMENT AND ITS EFFECT ON THE DEVELOPMENT OF STRESS IN THIN FILMS DURING DEPOSITION FROM THE VAPOR PHASE

D. Goyal and A.H. King

Department of Materials Science and Engineering
State University of New York,
Stony Brook   NY   11794-2275

Abstract

Transmission electron microscope observations of the structures of single-phase metallic thin films have been made.   Films prepared by sputtering and by thermal evaporation have been used and the structures have been evaluated quantitatively as a function of film thickness.   It is found that the grain size in all films increases linearly with the film thickness, indicating that grain growth occurs during film deposition.   It is demonstrated that grain growth acts effectively as a means of stress relief in the films.

Microstructural Science for Thin Film
Metallizations in Electronics Applications
Edited by J. Sanchez, D.A. Smith and N. DeLanerolle
The Minerals, Metals & Materials Society, 1988

Introduction

The mechanical and electrical properties of thin films can be substantially affected by their structures, and it is well recognized that a thin metallic film can be amorphous, polycrystalline or single crystal (epitaxial), depending upon the deposition process and composition. In particular the deposition rate and the substrate temperature have important influences upon the eventual structure. It has variously been suggested that the grain structure of a polycrystalline film is determined by the nucleation density or by recrystallization effects, both of which are temperature dependent. We have conducted a series of experiments upon thin films deposited at constant rates using thermal evaporation or ion sputtering, in order to assess what mechanisms are responsible for the development of their final structures. It will also be shown that the mechanisms that determine the film structures also have important effects upon the stresses developed in the films during deposition.

Experimental

In the present studies, chromium, nickel and tungsten films of varying thicknesses were deposited onto silicon single crystal wafers of (111) and (100) surface orientations. The substrates were cleaned only with non-corrosive solvents before use, so they were coated with native oxide layers. The substrates were nominally at room temperature for all depositions. Chromium and nickel films were deposited by thermal evaporation while tungsten films were deposited by ion sputtering. A background pressure of 10^{-6} torr was obtained before deposition commenced in all cases. The pressure during deposition typically rose to 10^{-4} torr. A deposition rate of 0.15nm per second was used for all of the coatings, and a range of coating thicknesses from 5 to 500nm was prepared.

Transmission electron microscopy specimens were prepared in two ways: the first is a modification of the apparatus suggested by Booker and Stickler [1]. Disc specimens were trepanned from the whole wafers and chemical jet polished from the silicon side, in order to yield thin specimens consisting of the substrate and the coating, which can be inspected in plan view. The second type of specimen preparation used ion milling rather than chemical polishing, and was used to form cross-sectional specimens, using the technique of Bravman and Sinclair [2].

The average grain size for each film was determined from plan view transmission electron micrographs using the mean linear intercept method. These calculations were based on values obtained from several readings, using a line length of 1 cm on micrographs taken at 100,000X magnification.

Results

For every system that we studied, it was found that the grain size that we measured was dependent upon the thickness of the film that was used. The grain sizes increased linearly with the film thickness, as is shown in Fig.1 and Fig.2, which represent thermally evaporated and sputtered films, respectively. It should be noted that for this experiment, the mean grain diameter was always much less than the film thickness, so that effects such as the Mullins limit on grain size [3] are not important. The cross sectional micrographs indicate that the films are columnar grained, and that the grains are parallel sided, as shown in Fig.3. It is clear that the Moire fringes sometimes observed in plan view micrographs result from either the projected overlap of adjacent columns, or the existence of small angle grain boundaries in the columnar grains. They do not indicate that the structure of the film is equiaxed.

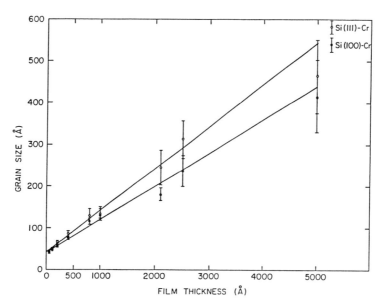

Figure 1a - Grain size as a function of film thickness for thermally evaporated chromium on (111) and (100) silicon substrates.

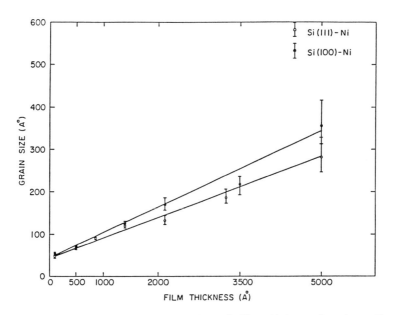

Figure 1b - Grain size as a function of film thickness for thermally evaporated nickel on (111) and (100) silicon substrates.

141

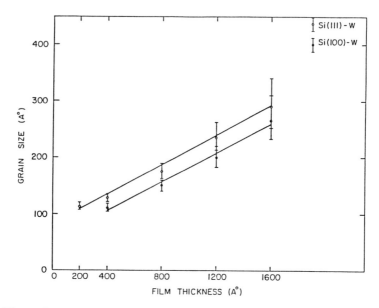

Figure 2 - Grain size as a function of film thickness for sputtered
tungsten on (111) and (100) silicon substrates.

The grain size results clearly indicate that grain growth occurs during the
deposition of thin films either by thermal evaporation or by ion sputtering. We
may extrapolate our results back to find the grain size at zero film thickness,
which may be interpreted as the mean separation between film nuclei. When we
do this, we notice that the nucleation density for thermally evaporated films is
identical for both (100) and (111) substrates, although the grain growth rates are
different. The mean separation between nuclei appears to be approximately 3nm,
which corresponds to the average migration distance for an adatom diffusing on
the surface of the substrate. On the other hand, for the sputtered films, the
growth rates are identical for (100) and (111) substrates, but the nucleation densities are different.

Discussion

Grain growth of the type found in this study has also been observed by
several other authors [4-7], although all of those studies have been carried out at
rather larger film thicknesses than in this case.

The causes of the grain growth during thin film depostion are not immediately apparent, since grain growth is a thermally activated process, and we
were using nominally room temperature substrates and high melting point
materials. However, there are two sources of heat available to the specimen
during the deposition of a thin film: The first is the enthalpy contained in the
evaporated material, including the latent heat of condensation of the film. The
second source of heat is radiant heat from the material source, operating at high
temperature, a few centimeters from the substrate. During each film deposition
we monitored the substrate temperature by means of a small K-type thermocouple
in contact with the back side of the substrate wafer. For thermal evaporations,
the temperature rise was found to be about 0.2°/nm for chromium and 0.4°/nm

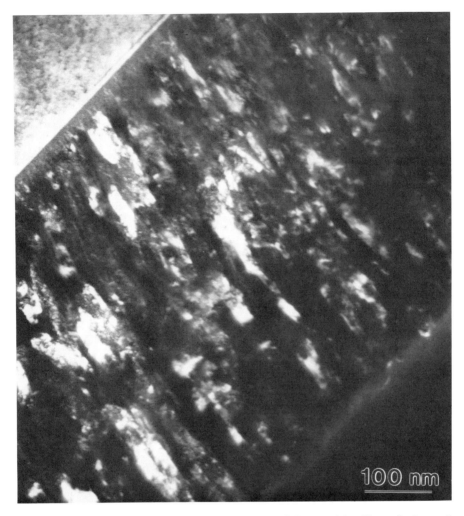

Figure 3 - Cross-sectional TEM image of a 500nm thick film of thermally evaporated chromium on a (100) silicon substrate, showing the columnar grain structure.

for nickel. The temperature rise for the sputtered tungsten specimens was approximately $0.1°/nm$. These results are consistent with a significant enthalpic contribution to the temperature rise, since the evaporation source for chromium is necessarily hotter than that for nickel, and so provides a greater radiant heat contribution, but the temperature rise for the nickel films was found to be larger. Since our thermocouples were placed on the backs of the substrates, we may infer that the temperature rise in the film itself is much larger than that which we measured, and indeed, a simple calculation assuming no conductive or radiative heat loss suggests that a temperature rise of a few hundred degrees per nanometer of deposited film can be expected.

The differences between the grain growth rates on (100) and (111) substrates are difficult to explain, but they must be considered to be real, since a statistical analysis reveals that the slopes of the lines are different at the 95% confidence level for both of the materials used here. It is particularly puzzling that the faster grain growth rate occurs on (100) substrates for nickel but on (111) for chromium. The difference in the nickel case may be attributed to the different thicknesses of the native oxides that are found on the two different orientations, because the oxide provides an effective thermal barrier, preventing conductive heat losses to the substrate. With this reasoning, the thicker oxide layer on the (111) substrates should result in higher temperatures being reached, and thus in faster grain growth, as observed. For the chromium case it is possible that a reaction occurs between the native silicon oxide and the chromium, to form chromium oxide. If this is an exothermic reaction at the effective interface temperature it may act to further increase the temperature and hence the grain growth rate, and the heat source would be more extensive for the case of a thicker oxide layer. We note that a fine-grained layer is observed at the film-substrate interface in Fig.3, but conclusive diffraction evidence has not yet been obtained to establish the structure of the layer.

The differences between the experiments on different substrate orientations for the sputtered films are more difficult to rationalize. It is conceivable that the grain growth rates are identical because the sputtering process acts to clean the native oxide from the surface of the substrate, so the heat losses become identical. The shift of the two curves relative to each other could result from the longer time required to remove the thicker oxide layer on the (111) substrates. Clearly, however, the nucleation process for sputtered films is rather different from that of thermally evaporated films, and further experimentation on this question is called for. In this regard, we will be undertaking studies using films with controlled thermally oxidised layers.

Since the grain growth of the specimens is a thermally activated process, it is to be expected that the same process will lead to the relaxation of grown-in film stresses, whatever their source might be. Typically, it is found that evaporated films are formed in a state of tensile stress, while sputtered films appear to be compressively loaded [8], but these statements refer to the average state of stress for the film as a whole. When films are formed with a tensile internal stress for reasons such as constrained thermal contraction, each new layer of material added to the film will relieve the stress in the underlying layer, by means of recrystallizing it or permitting grain growth, but the new layer will be formed with the intrinsic stress present. This reasoning leads us to conclude that the grown-in stress is not homogeneous through the thin film, but is largest at the top surface and smallest (or even conceivably reversed) at the bottom surface. Evidence for such stress distributions is found when thin films delaminate from their substrates as shown in Figs. 4 and 5, which show evaporated films (tensile surface stress) curling upward, and sputtered films (compressive surface stress) curling downward, respectively. It is found that the average stresses in the films are smaller in those cases where the grain size is larger [8], clearly suggesting that the stress relaxation and grain growth are linked.

The actual causes of the stresses in films formed from the vapor phase are not well understood, particularly for the case of sputtered films. In this study, we have shown that whatever processes lead to the formation of tensile stresses in evaporated films and compressive stresses in sputtered films, they must take place at the surface of the film, probably during the solidification process or very shortly thereafter. If this were not true, then the auto-annealing process would not lead to the stress distributions which are graphically demonstrated in Figs. 4 and 5.

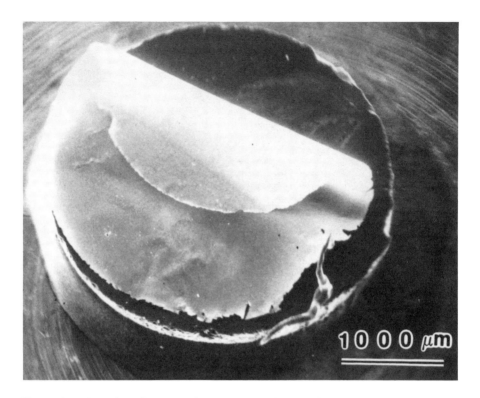

Figure 4 - Scanning electron micrograph showing curling of a 320nm evaporated nickel film on a (100) silicon substrate. The sign of the curl indicates that the top surface is in a state of higher tensile stress than the bottom surface of the film.

It is interesting to note that recrystallization typically results in an increase in density, because of the reduction in grain boundary free volume. This effect should give rise to an increasing tensile stress, or a decreasing compressive stress as recrystallization proceeds. The magnitude of this effect, however, would appear to be small compared with the stresses that are relieved by the formation of new stress-free grains. Indeed if this were not the case, then the increasing stress resulting from the reduction of grain boundary area would reduce the driving force available for grain boundary migration. When the overall stress is reduced, the driving force is increased through the loss of strain energy in addition to grain boundary energy, and this increase might contribute to the linear kinetics, which are not observed in isothermal recrystallization of stress-free materials.

145

Figure 5 - Scanning electron micrograph showing curling of a 160nm sputtered tungsten film on a (100) silicon substrate. The sign of the curl indicates that the top surface is in a state of higher compressive stress than the bottom surface of the film.

Conclusions

The intrinsic stresses and microstructures of thin films deposited onto substrates are considerably modified by an auto-annealing process facilitated by the deposition of heat with the material itself. Models for the generation of film stresses need to take this into account, and models for the grain growth during film formation also need to be constructed without making any assumption of isothermal conditions.

Acknowledgment

This work was supported by the Army Research Office, under contract number DAAL03-87-K-0049.

References

1.   G.R. Booker and R. Stickler, Brit. J. Appl. Phys. 13, 446 (1962).
2.   J.C. Bravman and R. Sinclair, J. Electr. Micr. Tech., 1, 53 (1984).
3.   W.W. Mullins, Acta Met. 6, 414 (1958).
4.   K.L. Chopra, Thin Film Phenomena (McGraw Hill Book Company, New York, 1969), p.183.
5.   A.S. Nowick and S. Mader, in Basic Problems in Thin Film Physics, edited by R. Niedermayer and H. Mayer (Vandenhoeck & Ruprecht, Goettingen, 1969), p. 212.
6.   E.I. Alessandrini, M.O. Aboelfotoh, R.B. Laibowitz and J.A. Lacey in Thin Films: The Relationship of Structure to Properties, edited by C.R. Aita and K.S. Sreeharsha (Mater. Res. Soc. Proc. 47, Pittsburgh, PA 1985), pp. 27-34.
7.   R. Berger and H.K. Pulker, in Thin Film Technologies, edited by J. Roland Jacobson (Proceedings of S.P.I.E. 401, Bellingham, WA, 1983), pp. 69-72.
8.   W. Ng, MS Thesis, State University of New York at Stony Brook, 1986.

Session III:

MICROSTRUCTURE -> PROPERTIES RELATIONSHIPS, FAILURE MECHANISMS

Session Chairman
John Sanchez
Lawrence Berkeley Laboratory
University of California
Berkeley, California

Stresses and Deformation Mechanisms in Thin Films

M. F. Doerner

IBM General Products Division

San Jose

Stresses that develop in thin films can be detrimental to the reliability of electronic devices. It is important to understand
the origin of these stresses and the consequent deformation that
can result. This review highlights the role of thin film microstructure in the development of stress and in deformation processes in thin films. A more complete discussion of this subject
can be found in a previous review(1). The three major sources of
stress development in thin films are 1) thermal stresses due to
differential thermal expansion of the film and substrate, 2)
stresses arising from lattice mismatch in the case of epitaxial
films and 3) intinsic (or growth) stresses that result from non-
equilibrium growth of the film. In this review, some models are
presented to describe the development of stresses due to non-
equilibrium growth of a film. The rates of stress generation due
to microstructural changes in a film are calculated. The microstructure of thin films is also an important consideration in the
operation of stress relaxation mechanisms. The relationship of
the microstructure and constraint of the substrate to deformation
in thin films is also discussed.

Microstructural Science for Thin Film
Metallizations in Electronics Applications
Edited by J. Sanchez, D.A. Smith and N. DeLanerolle
The Minerals, Metals & Materials Society, 1988

Stresses

Stresses in thin films on substrates are produced by processes
that would cause the dimensions of the film to change if it were
not attached to the substrate. In the case where the lateral dimensions of the film are large compared to the thickness, the
stress that results is pure biaxial tension or compression in the
plane of the film. Shear or normal stresses act on the
film/substrate interface only very near the edges of the film.
The edge stresses are important to the initiation of delamination. For patterned films where the lateral dimensions are
comparable to the thickness, the edge stresses become more important and the biaxial stress in the film is reduced. Aleck's(2)
solution for the thermal stresses in a plate with a clamped
boundary has been corrected by Blech and Levi(3) to give the
stress distribution at the edge of the plate. A complete solution for the stress distribution at the edge of the film for the
case of finite elastic rigidity of the substrate, however, is not
available.

Grain growth

The first mechanism to be considered in the development of intrinsic stresses is that of grain growth. As discussed by
Chaudhari(4), the elimination of grain boundaries that are less
dense than the crystal lattice results in a densification of the
film with an associated increase in the tensile stress. The rate
of stress generation from grain growth can be calculated by modification of an existing kinetic grain growth solution(5,6) to
include the effect of stress on the driving force. Sample calculations for a Ni thin film show that at room temperature the rate
of stress generation from grain growth is not significant except
for extremely fine grain sizes.

Vacancy annihilation

Due to the rapid growth rates encountered in thin film deposition, an excess vacancy concentration is common. Annihilation of
vacancies at the interfaces and internal grain boundaries can
produce stresses. Since the partial molar volume of a vacancy is
usually less than that of an atom, annihilation of a vacancy at
the surface will cause the film stress to become compressive.
Annihilation of vacancies at the grain boundaries perpendicular
to the interface, however, will generate a tensile stress since
elastic stretching of the grains will be required to close the
gap formed by the depositing vacancies. The rate of stress generation from vacancy annihilation can be calculated by solving
the diffusion problem of vacancy migration to the grain boundaries. A calculation for vacancy diffusion in Ni shows the rate
of stress generation from this mechanism to be slow below about
500K.

150

Void Closure

Another model for the evolution of stresses in thin films involves the filling of grain boundary voids due to diffusion of
atoms along the grain boundaries. Here, as in the case of vacancy annihilation, the gap formed at the grain boundaries must
be accomodated by elastic strain in the grains. The kinetics of
this process are calculated using the void growth solution of
Speight and Beere(7). For small voids that are evenly distributed along the grain boundaries, stress generation can proceed
at reasonable rates at room temperature.

The stresses that arise from non-equilibrium growth of a film are
often maximum during the early stages of film growth just following the coalescence of individual crystallites into a continuous
film. The mechanism for stress generation during coalescence may
be similar to that of void closure. As the channels between
crystallites fill up during coalescence, some of the gap is
accomodated by elastic strain in the crystallites. Grain growth
can also be significant during this stage of film growth and may
also contribute to the large stresses that are observed.

Deformation Mechanisms

Dislocation processes

The flow stress of a thin film is generally observed to be higher
than that of the corresponding bulk material. In addition, the
strength of thin films is often observed to be thickness dependent with thinner films ususally having higher strengths. The increased strength of thin films can be partially explained by the
increased resistance to dislocation motion due to high dislocation densities, high impurity concentrations and small grain
sizes. For Al films(8), the change in strength with film thickness could not be explained by the grain size assuming the Hall-
Petch coefficients for bulk Al. The presence of the stiffer
substrate and oxide phases is believed to also be important to
the increased resistance to dislocation motion in thin films.

Grain boundary diffusion

Due to the unique columnar microstructure present in thin films
and to the constraint of the substrate, grain boundary diffusion
has limited potential for stress relaxation in thin films. For a
thin film under a biaxial compressive stress, there is a driving
force for diffusion of atoms from grain boundaries perpendicular
to the film/substrate interface to grain boundaries lying parallel to the interface. In a columnar microstructure, the lack of
grain boundaries parallel to the film/substrate interface requires deposition or removal of atoms to occur at the
film/substrate interface or at the free surface. Since in the
case of metals, an oxide phase is often present at the free surface, diffusion along this interface is difficult. The lack of
adequate sources and sinks for atoms along parallel boundaries
limits the rate of stress relaxation from grain boundary diffusion. The observation of inhomogeneous deformation in thin

films (hillocks) is probably a result of this limited diffusion. Diffusion proceeds at local sites in the microstructure where parallel boundaries may be present or where cracks exist at the interfaces.

The constraint of the substrate also limits the potential for stress relaxation from grain boundary diffusion(9). Although the stress at the grain boundary can be relaxed, the constraint of the substrate prevents the stresses at the interior of the grain from being totally relaxed. The remaining stress in the grain is the same as that for an isolated island on a substrate. The suppression of grain boundary diffusion is evident in data obtained by Murakami and Kuan(10) for Pb films.

References

1. M.F. Doerner and W.D. Nix, to be published in CRC Critical Reviews in Solid State and Materials Science.

2. B.J Aleck, J. Appl. Mech. 16 (1949) 118.

3. I.A. Blech and A.A. Levi, J. Appl. Mech. 48 (1981) 442.

4. P. Chaudhari, J. Vac. Sci. Technol. 9 (1972) 520.

5. P.G. Shewmon, Transformations in Metals, Mc-Graw Hill, NY (1969) p. 108

6. D.A Smith, C.M.F. Rae and C. R. M. Grovenor in Grain-Boundary Structure and Kinetics, ASM, Metals Park, Ohio (1980) p.337.

7. M.V. Speight and W. Beere, Met. Sci. 9 (1975) 190.

8. M.F, Doerner,D.S. Gardner and W.D. Nix, J. Mater. Res. 1 (1986) 845.

9. M. S. Jackson and C.-Y. Li, Acta Metall. 30 (1982) 1993.

10. M. Murakami and T.S. Kuan, Thin Solid Films 66 (1980) 381.

THIN FILM ADHESION

R. M. Cannon, R. M. Fisher

Center For Advanced Materials
Lawrence Berkeley Laboratory
University of California
Berkeley, CA 94720

Abstract

Film adherence is assessed from a fracture perspective emphasizing microstructure and composition effects. Analyses of dissimilar interface fracture energies elucidate large chemistry effects from prior segregant and from environments. Often the fracture energy derives largely from plasticity within metal members and crack-microstructure interactions which, in turn, depend on the chemical debonding. The adhesion depends upon these fracture resistance aspects plus residual and applied stresses driving delamination. Studies of model systems affirm fracture mechanics predictions that residually stressed films may delaminate after attaining a critical thickness. The fracture modes and probabilities depend upon precursor flaws, geometrical discontinuities and, whether the stress is tensile or compressive. Often high stresses developed during growth which are very sensitive to growth conditions and resultant microstructures are dominant. Novel methods to degrade or enhance fracture energy and film delamination resistance are used to establish correlations.

Microstructural Science for Thin Film
Metallizations in Electronics Applications
Edited by J. Sanchez, D.A. Smith and N. DeLanerolle
The Minerals, Metals & Materials Society, 1988

MORPHOLOGY AND MICROSTRUCTURE OF LPCVD TUNGSTEN FILMS

John C. Bravman, David C. Paine

Materials Science and Engineering
Stanford University
Stanford, California 94305

Abstract

For VLSI metallizations interest has recently focused on refractory metal systems in general and in particular on tungsten deposited by low pressure chemical vapor deposition (LPCVD). This technology is attractive for submicron structures because of its low resistance, ohmic behavior for both n- and p- type silicon, its high melting point, and its conformal deposition in high aspect ratio vias. Of special interest is its unique selective deposition character, in which films of tungsten nucleate and grow on exposed silicon but not on oxide or nitride dielectrics. We describe here our TEM investigation of two related features: the formation of a metastable, high resistivity allotrope of tungsten, known as β-tungsten, and the growth of filamentary tunnels in the silicon and beneath the tungsten which terminate in a tungsten-rich particle. We have studied films deposited in two different types of reactors operated under a variety of common deposition conditions; temperature, gas flows, and silicon surface treatment were the experimental variables. Both patterned and unpatterned wafers were used. Conditions which favor the deposition of β-tungsten will be discussed, as will significant differences between the blanket deposits and those found on patterned wafers.

Microstructural Science for Thin Film
Metallizations in Electronics Applications
Edited by J. Sanchez, D.A. Smith and N. DeLanerolle
The Minerals, Metals & Materials Society, 1988

MICROSTRUCTURAL CHARACTERIZATION OF

ALUMINUM-1% SILICON FOR INTEGRATED CIRCUIT APPLICATIONS

Bryan M. Tracy, Paul W. Davies, Dave Fanger,
and Pam Gartman

Intel Corporation
3065 Bowers Ave, M/S SC2-24
Santa Clara, CA 95051

Abstract

The grain size of sputter-deposited Al-1%Si films has been found to obey the lognormal distribution. The grain size distribution was determined as a function of substrate temperature, deposition equipment type, thermal processing and film composition. In each case, lognormal behavior was observed except under conditions where the grain size exceeded the film thickness. The development and usefulness of the lognormal distribution is presented and discussed.

Microstructural Science for Thin Film
Metallizations in Electronics Applications
Edited by J. Sanchez, D.A. Smith and N. DeLanerolle
The Minerals, Metals & Materials Society, 1988

Introduction

The purpose of this work is to better understand the changes in microstructure of Al-1%Si thin films as a result of typical integrated circuit thermal processing. The work presented here will highlight our results from materials studies on approximately one hundred such films. The primary analytical technique used in this work was Transmission Electron Microscopy (TEM). In the course of the analysis, the log normal size distribution was found to better represent the aluminum grain size than the conventional Gaussian distribution.

There are several advantages of using the lognormal distribution to describe thin-film grain size statistics. Firstly, and perhaps most simply, the lognormal distribution is preferred to the Gaussian distribution because, it better describes the film, i.e., the number of small grains is larger than would be predicted by normal statistics.

Secondly, because the entire distribution is represented by a line, it is quite easy to graphically examine and compare the effect of any processing parameter, i.e., annealing which might change the grain size distribution.

As a final point, one of the major goals in the area of semiconductor reliability is to determine which microstructural parameter lead to superior electromigration performance. Fraser and co-authors (1) have elegantly showed that both the average grain size and the grain size distribution directly control the Mean Time To Failure (MTTF) of aluminum thin films. it is hoped than an improved description of the size distribution will assist in the development of predictive models like that of Fraser (1) leading to semiconductor devices with improved reliability.

Experimental

The grain size distribution for all films was evaluated by Horizontal Transmission Electron Microscopy (HTEM) using the general sample preparation method of Marcus and Sheng (2). Unpatterned wafers, with a nominal Al-1% Si thickness of 1.0 microns deposited on 1000A of SiO_2, were back-ground to 100 microns, dimpled to 50 microns, and then two-step etched in HNO_3 and HF (4:1) to remove the remaining silicon. The samples were then ion milled at the shallowest possible angle (12°) for the shortest possible time in order to give clear images when examined in a JEOL 1200EX at 120 KeV. The importance of using a shallow milling angle and brief milling time cannot be overstated. Aluminum is a very difficult material to successfully ion mill and is prone to "cone formation" and differential crystallographic sputtering. The best samples were not ion milled to perforation, but rather back-thinned to approximately 0.5 microns.

As will be shown, this technique poses no threat to the evaluation of aluminum grain size, but it does have the disadvantage of removing approximately the bottom half thickness of the film. Silicon precipitation has been found to occur preferentially in this region (4). Thus, for a full characterization of the film, some combination of Cross Sectional Transmission Electron Microscopy (XTEM) and a depth profiling technique such as Secondary Ion Mass Spectroscopy (SIMS) is required. An alternative for microstructural evaluation would be the use of High Voltage Electron Microscopy (HVEM) which is able to clearly image at 1 MeV the full thru-thickness of a 1.0 micron aluminum film without any ion milling.

The grain size distribution was measured directly from the TEM micrographs by tracing each grain onto a digitizing tablet. The grain size distribution is then calculated from the individual measurements using a

semi-automated image analysis software program called VIAS^{TM}. The
lognormal plots were created using a graphics package known as RS1.

Results And Discussion

1) Presentation of Typical Results

The microstructure of an as-deposited aluminum film is shown in Figure
1. The grain size histogram, Figure 2, shows a poor fit to the Gaussian
Distribution.

When the grain size is transformed to log grain size, a much better
fit, indicating a lognormal size distribution, is obtained. Please refer
to Figure 3. The immediate utility of the lognormal presentation is seen
when the data is plotted on probability paper (see Figure 4). Straight
line behavior is observed and the coefficient of variation, which is
effectively the standard deviation normalized by the mean is calculated to
be 0.4.

500nm

Figure 1

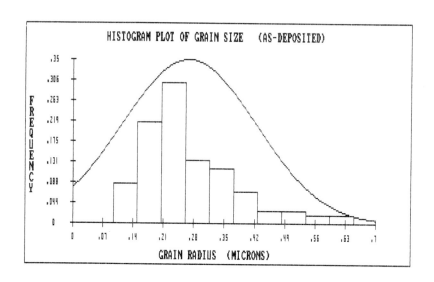

Figure 2

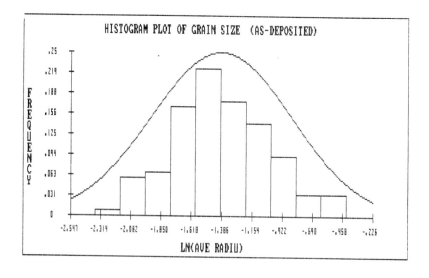

Figure 3

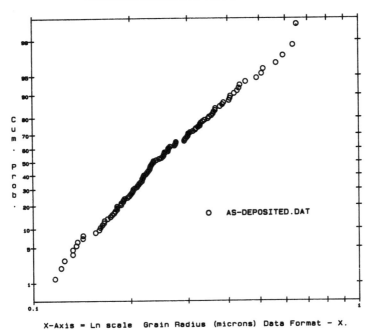

Figure 4

2) Test of Goodness of Fit (Gaussian vs Lognormal)

As mentioned above, there are several indications that the grain size of sputter-deposited Al-1% Si follows the lognormal size distribution. These indicators are summarized below in order of increasing importance.

a) Poor fit to Gaussian distribution.

b) Better fit to Gaussian distribution when the log of the grain size is plotted.

c) Straight line behavior of the log of the size distributions when plotted on probability paper.

d) Failure of approximately half of the distributions to pass the Kologorov-Smirnov test (5) for the normal distribution.

3. Effect of Substrate Temperature

One of the more common variables during the sputter-deposition of Al-1% Si is the substrate temperature. In order to examine the effect of substrate temperature on grain size, a series of wafers was deposited in a Varian 3180 at 50C increments. The range of substrate temperature was 50C to 350C. The size distribution results are presented in Figure 5.

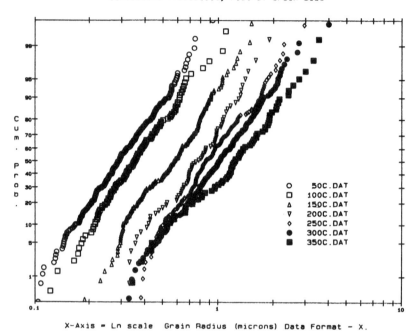

Cumulative Probability Plot of Grain Size

X-Axis = Ln scale   Grain Radius (microns) Data Format - X.

Figure 5

Once again the fit to the lognormal distribution is good, and the coefficients of variations of all the distributions are quite close to 0.4, regardless of median size (substrate temperature). As expected, we see that median grain size increases with substrate temperature. Some non-linearity is seen at the higher deposition temperatures. At approximately 150C, the median grain size equals the film thickness. It is suggested that at this point some type of modified two dimensional grain growth occurs because the films is restrained from growth in the Z direction. Such a growth process may give rise to the observed non-linearity; more work is required to elucidate this point.

4. Effect of Deposition System

In the course of this study, we had the opportunity to evaluate Al-1% Si films from several different deposition systems. In this paper we present grain size results from the Perkin Elmer, Varian 3180, Anelva, and Gartek sputter systems. The grain size distributions are shown in Figure 6. It is notable that despite the design differences between these systems, each film follows the lognormal distribution. The Varian film has the largest median grain size because a substrate temperature of 250C is employed. The coefficients of variations for each distribution is within +/- 0.15 of the average value of 0.45.

In fact, we have found the median grain size to be a very sensitive function of the actual wafer temperature during deposition. Please refer to Figure 5. Thus, any deposition parameter which effects the wafer temperature such as deposition power, substrate thickness, and the actual number of wafers processed, will change the median grain size. The differences in the size distributions seen from these four deposition systems is attributed to variations in the wafer temperature.

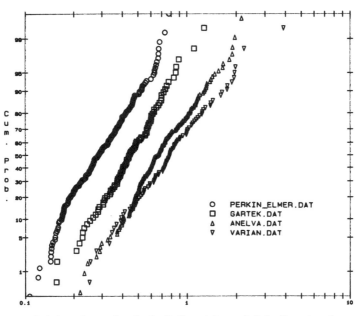

Cumulative Probability Plot of Grain Size

X-Axis = Ln scale   Grain Radius (microns)   Data Format - X.

Figure 6

5. Effect of Annealing

Considerable grain growth occurs during annealing in forming gas at 400C for 20 minutes. The median grain diameter increased from about 0.5 microns to over 5.0 microns. Interestingly, even the alloyed distribution, Figure 7, is reasonably well fitted to the lognormal plot, however, the dispersion (coefficient of variation) is increased from 0.40 to 0.60.

We suggest that the film thickness effects the grain growth process during anneal, because the final grain size, when viewed in projection by the TEM, is a factor of 10 larger than the nominal thickness. Thus the grains take on the shape of short, wide-based columns. Some alteration in the final grain size distribution is expected because grain growth in the "Z" direction is prohibited. This effect needs further study and clarification.

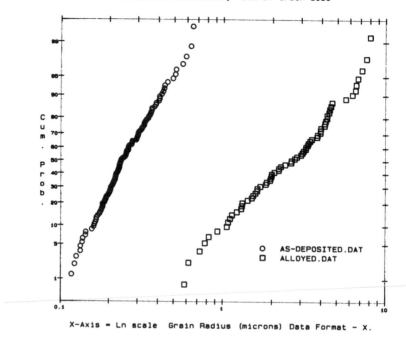

Cumulative Probability Plot of Grain Size

C
u
m
.
P
r
o
b
.

O   AS-DEPOSITED.DAT
□   ALLOYED.DAT

X-Axis = Ln scale   Grain Radius (microns) Data Format - X.

Figure 7

6.   Effect of Composition

Though we have a great deal less experience with Al-Cu films, our preliminary work shows that the lognormal grain size distribution is also followed by an Al-0.5%Cu film. The microstructure, which shows numerous CuAl$_2$ precipitates, is presented in Figure 8. It is interesting to note that the grain size distribution, Figure 9, is still lognormal despite the presence of these precipitates which can be reasonably expected to effect the nucleation and growth of the film.

Figure 8

Cumulative Probability Plot of Grain Size

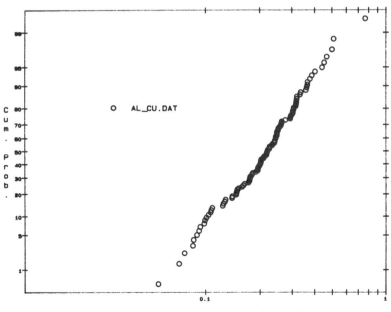

○   AL_CU.DAT

X-Axis = Ln scale   Grain Radius (microns) Data Format - X.

Figure 9

7. Effect of Microstructure

Several films were cross sectioned to evaluate the through-thickness microstructure of Al-1% Si. This was done to insure that the grain structure did not change with thickness as has been reported for thick polysilicon (1). Were this the case, our back-thinning procedure would result in an erroneous grain size measurement. A cross section of a an Al-4%Si film shows a blocky-columnar microstructure in which most grains extend across the full thickness of the film. Please see Figure 10. Back-thinning such a film should not result in a serious mis-measurement of the aluminum grain size. Because the silicon content is quite high, it is easy to locate precipitates near the bottom of the film.

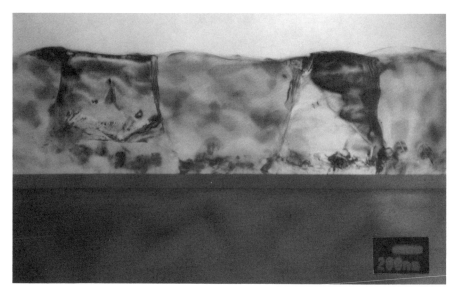

Figure 10

Conclusions

The usefulness of the lognormal distribution has been demonstrated for the characterization of sputter-deposited Al-1% Si. This lognormal behavior was found to be general, except in the case of elevated substrate temperature depositions, in which the median grain size exceeded the film thickness. It is suggested that when grain growth becomes semi-two dimensional, as in this case, some non-linearity in the lognormal distribution will be observed. Four different commercially available sputter-deposition systems yielded films which were all lognormal distributed. Preliminary experiments indicate that Al-0.5%Cu alloys also follow the lognormal distribution. It is hoped that predictive models for electromigration performance which use microstructural parameters such as texture and grain size distribution will benefit from the present work.

Acknowledgments

We would like to thank Yvonne Cummings and Joy Hilario for making many of the grain size distributions measurements and for the typing of the manuscript. Ron Wu was responsible for writing the data transfer and cumulative probability computer programs. His help was essential and greatly appreciated.

References

[1] D. B. Fraser, S. Vaidya, A. K. Sinha, "Electromigration Resistance of Fine-Line Aluminum for VLSI Applications, IEEE, (1980), 165-170.

[2] R. B. Marcus and T. T. Sheng, Transmission Electron Microscopy of VLSI Circuits and Devices (New York, John Wiley & Sons, 1983), 15-31.

[3] C. C. Chaing, T. T. Sheng, D. V. Sweeney, and D. B. Fraser, "Silicon Depth Profile and Contaminants in Si-Doped Al Film", Journal of Applied Physics, Vol 47, No. 5, May 1976, 1790-1794.

[4] N. A. Haroun, "Grain Size Statistics", Journal of Material Science, 16 (1981) 2257-2262.

[5] J. Zar, "Biostatistical Analysis", Prentice Hall, 1974, 81.

ACTIVATION ENERGIES ASSOCIATED WITH CURRENT NOISE

OF THIN METAL FILMS

J. G. Cottle & T. M. Chen

Department of Electrical Engineering
University of South Florida
Tampa, FL, 33620

ABSTRACT

Current noise measurements of thin metal films may be used as a valuable tool for studying their material properties. The noise voltage spectra are sensitive to microscopic processes which affect conduction and govern effects such as electromigration. This paper reviews methods for determining activation energies associated with the $1/f$ and $1/f^2$ current noise components in thin metal films. The activation energies are close to those reported in the literature for grain boundary diffusion. Although the $1/f$ noise increases dramatically with the structural damage caused by electromigration damage, the $1/f^2$ noise component is more directly related to the electromigration process. Direct activation energy measurements based on the $1/f^2$ noise component's temperature dependence yield 0.59eV, 0.69eV and 0.74eV for Al(99.999%), Al-Si(2%) and Al-Cu(2%), respectively. Possible generation mechanisms of current noise and its relation to the electromigration process will be discussed.

Microstructural Science for Thin Film
Metallizations in Electronics Applications
Edited by J. Sanchez, D.A. Smith and N. DeLanerolle
The Minerals, Metals & Materials Society, 1988

Introduction

The thin metal films used as interconnections in VLSI must tolerate the adverse effects of higher and higher current densities if a given level of reliability is to be maintained while feature size is reduced. The most serious of these effects is electric field assisted mass transport (electromigration) along the length of the film which results in the formation of metallic accumulation (hillocks) and depletion (voids). In the extreme, two failure modes result: 1) a total depletion of metal from an area causing an open circuit or; 2) a short to adjacent interconnection.

Mass transport results from processes which govern the diffusion of vacancies along the grain boundaries of a conventionally sputtered or thermally evaporated thin film. The application of a direct electric current imparts a preferred direction to the net interaction between metal atoms and vacancies. The force on an ion, $\vec{F_i}$, due to an externally applied field, $\vec{\mathcal{E}}$, is given as

$$\vec{F_i} = e\vec{\mathcal{E}}(Z_0 - Z_e) = e\vec{\mathcal{E}}Z^* \tag{1}$$

where e represents the value of the electron charge and Z^* represents an effective ion valency which is the sum of valence of the metal ion, Z_0, and a drag term, Z_e, due to the force exerted through collisions of electrons with ions. If the drag force is large, the sign of Z^* is negative and the force results in a migration of metal atoms against the electric field or toward the positive terminal of the battery. The mass flow along the thin film interconnection may be expressed in terms of an ionic drift velocity as follows

$$\vec{v} = \frac{\vec{J_i}}{N} = \vec{j}\rho \left(\frac{eZ^*}{k_B T}\right) D_0 \, exp \left(\frac{-E_a}{k_B T}\right) \tag{2}$$

where $\vec{J_i}$ is the ionic flux, N is the density of metal ions, $\vec{j}$ is the current density, k_B is Boltzmann's constant, T is absolute temperature, D_0 is the infinite temperature diffusion constant, ρ is the resistivity of the sample and E_a is an activation energy associated with the diffusion process. E_a is an aggregate activation energy which may represent bulk, grain boundary, and surface diffusion processes. Schreiber has summarized the following five contributing mechanisms which govern mass transport in metal films, with their respective activation energies for pure aluminum [1].

1. Bulk electromigration ($E_b = 1.4eV$)

2. Grain boundary electromigration ($E_g = 0.4 \rightarrow 0.5eV$)

3. Vacancy motion from grain boundaries into the bulk ($E_{g/b} \approx 0.62eV$)

4. Motion of vacancies from defects in the bulk ($E_d > 0.62eV$)

5. Oxide free surface electromigration ($E_s \approx 0.28eV$)

Most of the activation energies above were indirectly determined[1]. The commonly used experimental techniques for activation energy determination are dependent on total material flux and therefore are measurements based on the sum of all contributing mechanisms listed above. One of the most accurate measurements of total mass flow is obtained directly via the edge displacement technique, originally suggested by Blech and Kinsbron [3,4].

Industry has traditionally characterized the median time to failure (MTF) of an interconnection to test its resistance to the effects of electromigration. MTF is by its very nature a time consuming measurement made at accelerated conditions by using current densities and temperatures much higher than those encountered by the film in normal use. MTF measurement determines the time t_{50} at which 50% of a large number of identical test samples fail due to electromigration. This time is usually fit to the following empirical equation, developed by Black [5].

$$t_{50} = MTF = Aj^{-n}exp\left(\frac{E_1}{k_B T}\right) \tag{3}$$

where A and n are constants, j represents current density, and E_1 is an activation energy associated with the electromigration process. The activation energy values for aluminum reported in the literature range from 0.4eV to 0.8eV and generally support the view that electromigration is a current density enhanced, grain boundary diffusion process [6].

Conventional MTF testing is both costly and time consuming and therefore delays the incorporation of newly developed metallurgies in integrated circuits. In addition, the conventional MTF test is

---

[1]E_b has been measured using nuclear relaxation [2].

destructive to the samples being tested. Industry is in need of a faster, non-destructive technique with which to characterize new metallurgies and to project the reliability of existing metal interconnections. Several other techniques have been proposed including Standard Wafer-Level Electromigration Accelerated Test (SWEAT) [7], Breakdown Energy of Metal (BEM) [8], Temperature Ramp Resistance Analysis (TRACE) [9], and assorted wafer-level techniques [10]. Noise measurements also show great promise as a quick non-destructive electromigration test [11,12,13], because the same microscopic processes which govern electromigration also affect conduction.

Current Noise of Thin Metal Films

In general, the noise voltage spectral density generated by an aluminum based thin metal film is the sum of a thermal and an excess (current) noise term and may be represented as

$$S_v(f) = 4k_B T R + \frac{K V^\beta}{f^\alpha} \qquad (4)$$

where R is the resistance of the film, K, β and α are constants and V is the applied d.c. bias voltage[2]. β and α are used to describe the dependence of the spectrum on applied voltage and frequency, respectively. For small volume films or films which show damage due to electromigration, the values of β and α are observed to be ≈ 2 and ≈ 1, respectively. Because $\alpha \approx 1$, we refer to this noise component as $1/f$ noise. K is dependent on film microstructure. In characterizing the excess noise term, two distinctly different frequency dependencies were noted for the the spectrum. When the d.c. biasing current is small and/or the film temperature is low, the measured noise spectra exhibit a $1/f$ frequency dependence, i.e., with $\alpha = 1$ in equation (4). As the temperature and/or bias current is increased however (such as those parameters commonly used in accelerated MTF testing), the value of α shifts from around 1 toward a value of 2. When present, the $1/f^2$ noise component dominates the entire low frequency portion of the noise spectrum and causes a large increase in the integrated, total noise voltage at low frequencies. There is a distinct current density threshold which stimulates the $1/f^2$ noise mechanism which decreases for an increase in film temperature. The differences in the characteristics of the $1/f$ and $1/f^2$ noise discussed below support the view that the components are due to different generation mechanisms.

Dutta and Horn have shown that the normalized $1/f$ noise (S_v/V^2) will have a maximum magnitude at a temperature, T_p, if an assumption is made that the noise arises from a superposition of thermally activated events and their corresponding time constants [14]. The temperature dependence of the noise may therefore be used to solve for an activation energy of the $1/f$ noise process, E_p, with

$$E_p \approx -k_B T_p \, ln(\omega \tau_0) \qquad (5)$$

where ω is the frequency of observation and τ_0 is a "typical" attempt time for the random process (assumed to be on the order of an inverse phonon frequency). The question, however, of exactly what process is governed by the thermally activated time constants is still not conclusively answered. The activation energy values obtained are sensitive to the assumed value of τ_0 [3] even though the spectrum predicted by Dutta and Horn [14] will remain $1/f$ for observed $\omega \ll \tau^{-1}$.

The $1/f^2$ noise is dominant at measurement parameters which are commonly used for MTF testing. The most profound difference between the $1/f$ noise and the $1/f^2$ noise is in their sensitivities to shifts in film temperature. Unlike the $1/f$ noise, the noise with a $1/f^2$ spectral slope is extremely sensitive to a shift in ambient temperature, with the spectrum represented by

$$S_{1/f^2}(f) = \frac{B}{f^2} j^m \, exp\left(\frac{-E_2}{k_B T}\right) \qquad (6)$$

where B is a constant, m is a number used to describe the behavior of the noise with current density, j, and E_2 is an activation energy associated with the $1/f^2$ noise process for a specific thin film interconnection. The values of E_2 determined directly from the temperature dependence of the $1/f^2$ noise agree well with the values of E_1 reported in the literature for grain boundary electromigration determined by a variety of measurement techniques [6,17].

Measurement Procedures and Results

Noise measurements were performed on packaged thin film samples using the system shown in Figure 1. The thin film samples were biased using sealed lead-acid batteries to eliminate power supply ripple

---

[2] Each term describes the asymptotic behavior of the spectrum.
[3] An order of magnitude change in τ_0 changes $E_p \approx 0.07$eV.

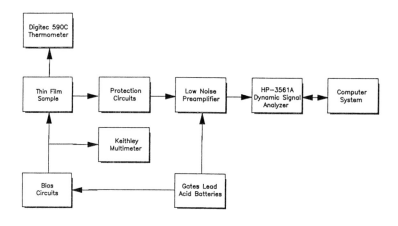

Figure 1: Block diagram of the noise measurement system.

components from the final results. The noise voltage of the thin film was capacitively coupled to a low noise amplifier with a voltage gain of 10000 (80dB). The amplified signal was analyzed with a Hewlett Packard 3561 dynamic signal analyzer which was initialized to perform 50 averages on a frequency span of 1Hz to 50 Hz with a flat-top spectral window. Data from the analyzer was transferred over the IEEE-488 instrumentation bus to a desktop computer for final analysis and plotting. The packaged parts (24-pin Ceramic DIPs) were placed in a heavy brass sample chamber which provided electrical shielding to avoid unwanted pickup from external sources. In addition, the brass offered a high heat capacity to maintain a relatively constant temperature (within a few degrees) so that the oven could be turned off during the acquisition of the noise signal. In the frequency range observed, shielding must be provided to guard against magnetic, electrical and vibrational disturbances in order to obtain an accurate spectral estimate of the noise generated by the thin film (typically on the order of $10^{-17}V^2/Hz$).

A plot of the noise spectrum generated by a particularly noisy thin film sample is shown in Figure 2. To determine the slope of the spectrum, single order, regressive fits are typically performed to the data between 1 and 10Hz. The intercept at 1Hz of the regression line is then determined and "normalized" by dividing by V^2 (the d.c. voltage across the sample) to account for the temperature dependence of the nominal film resistance value. Measurement parameter ranges (current density and temperature range) are selected so that predominantly one slope characteristic is observed over multiple temperatures. The low end limit in selecting current density is usually due to the sensitivity of the measurement system. At the current densities required for detection of the noise ($> 10^6 A/cm^2$), the temperature of the thin film increases significantly due to Joule heating. Joule heating must be taken into account in any analysis of the final measurement results.

$1/f$ Noise Measurements

Three types of samples exhibited noise voltage spectra which can be accurately described as having a $1/f$ dependence on frequency. $1/f$ noise was observed in thin films of small physical size ($1.5\mu m \times 0.8\mu m \times 600\mu m$), and larger volume films ($2.5\mu m \times 0.8\mu m \times 1250\mu m$) which had been damaged by electromigration. In addition, films damaged by mishandling in die-attach and wire-bonding exhibited current noise spectra which were predominantly $1/f$. Films damaged by electromigration show large magnitudes of $1/f$ noise when compared to undamaged interconnections.

There is no observable difference in the temperature or current dependencies of the $1/f$ noise gener-

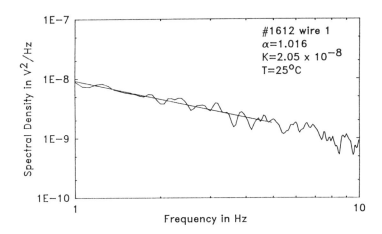

Figure 2: Current noise spectrum of a noisy thin metal film.

ated by films of small volume, films which show damage due to mishandling, or films with increased R due to electromigration. This fact implies that there is a common $1/f$ generation mechanism present in all the films, and no alteration of its descriptive parameters, α and β, with electromigration damage. For the case of $1/f$ noise, the values of α and β remain relatively constant ($0.9 \leq \alpha \leq 1.4$,$2.1 \leq \beta \leq 2.5$) for an undamaged film and one which has been has shown a change in resistance brought on by an acceler-ated electromigration test. Although α and β are relatively unaffected by the effects of electromigration, the magnitude of the $1/f$ noise increases dramatically. The value of K in equation (4) therefore is not a constant, but is very sensitive to the structural damage caused by electromigration. For a film with only a small resistance increase due to electromigration, K shows a large increase over that typically expected for an undamaged film of the same volume [12].

The experimental noise data for 10% $\Delta R/R_{initial}$, unpassivated Al(99.99%) and Al-Si(2%) films shown in Figures 3a and 3b show a temperature peak at 320°K and 360°K, respectively. If we assume an attempt time of 10^{-12} sec. (equation (5)), the activation energies corresponding to the data shown in Figure 3 turn out to be 0.71eV and 0.80eV for pure Al(99.99%) and Al-Si(2%), respectively. Such data is consistent with that reported by Koch et al. [13] [4], and is comparable to the activation energies reported for motion of vacancies from grain boundaries into the bulk [1]. The undamaged pure Al(99.999%) film data shown in Figure 3a yielded an activation energy, E_p, of 0.69eV corresponding to a peak $1/f$ noise temperature of 310°K.

A similar peak at 323°K was observed for a pure aluminum film with 10% ΔR due to electromigration damage. For this film, the absolute magnitude of the $1/f$ noise was approximately 3 orders of magnitude above that of the undamaged film (the curves have been altered to fit on the same axis), and a lower current density could be used to observe the $1/f$ noise. It is interesting to note that this large amount of structural damage (caused by electromigration) did not significantly alter the activation energy obtained from the normalized $1/f$ noise data.

Also observable in Figure 3a is an increase in the normalized $1/f$ noise for temperatures above 385°K. This trend is also reported by other researchers [13,14] and is most likely due to a departure of the noise spectrum frequency dependence from $\alpha = 1$ toward higher values of α. The increase in the value of α toward $\alpha = 2$ would accompany a non-equilibrium motion of aluminum atoms caused by electromigration and a departure from the j^2 dependence of the noise voltage spectrum.

---

[4]Koch et al. used $\tau_0 = 10^{-13}$ sec.

173

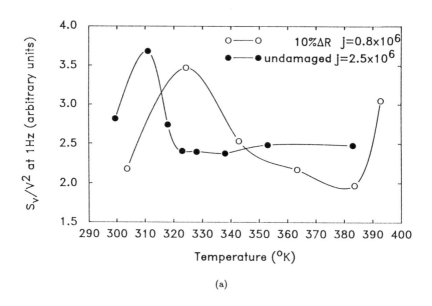

(a)

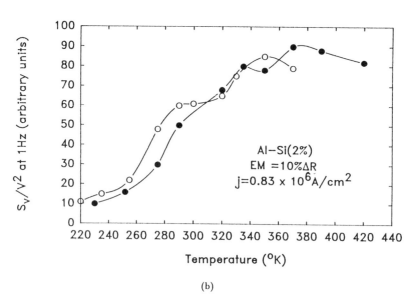

(b)

Figure 3: Normalized $1/f$ noise versus temperature (a) pure aluminum films (b) Al-Si(2%) films.

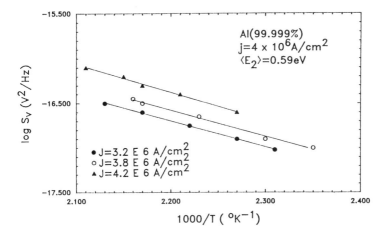

Figure 4: Exponential dependence of $1/f^2$ on film temperature for three Al(99.999%) films.

$1/f^2$ Noise Measurements

Although the $1/f^2$ noise is thought to be due to a non-equilibrium motion of aluminum atoms, larger volume films exhibit stable, repeatable noise spectrum measurements. For the short period of time needed to acquire the spectral data, no detectable change in overall film resistance is usually measurable. The smaller cross-sectional area films, or films showing a ΔR due to electromigration, do not exhibit stable $1/f^2$ noise.

Figure 4 illustrates the $1/f^2$ noise magnitude versus reciprocal temperature for three pure aluminum (99.999%) films at 1Hz. The current density was held constant at $4 \times 10^6 A/cm^2$ to obtain the data shown in the figure. The dependence is clearly exponential, as expressed by equation (6), with the slope yielding an average E_2 value of 0.59eV for this particular sample group. Similar measurements have been performed on films fabricated by varying material composition. From the spectral data for several Al-Si(2%) and Al-Cu(2%) films, average values for E_2 were found to be 0.69 and 0.74eV, respectively [5].

Preliminary measurements on films which are very thin (4000Å) at similar conditions have revealed difficulties in stimulating the $1/f^2$ noise behavior. For these films, no detectable $1/f^2$ noise occurs at 200°C until the current density is increased to around $6 \times 10^6 A/cm^2$. At this current density, the values of E_2 are typical of those reported for bulk diffusion (1.29eV).

Discussion

The activation energy values associated with current noise are very close to those governing the diffusion of vacancies in thin metal films. Since the current noise is sensitive to processes which affect conduction, it is a valuable tool for studying the properties of films used as interconnections in VLSI.

The temperature dependence of the $1/f$ noise typically yields activation energies somewhat higher than those reported in the literature for similar films tested using MTF accelerated testing. The $1/f$ process in thin metal films therefore, may be due to an equilibrium vacancy exchange between a grain boundary and the bulk of a grain. Films with smaller grain size and those damaged by accelerated testing show greatly increased magnitudes of $1/f$ noise indicating that the value of K is closely associated with the density of material defects in the film.

---

[5] The statistical distribution of E_2 values is currently being investigated.

The data describing the temperature and current density dependence of the noise spectrum data with a $1/f^2$ spectral slope strongly suggest that the observed fluctuation is linked to the atomic mass flow caused by electromigration. Indeed, it may be shown that $1/f^2$ noise arises when considering the scattering which accompanies a slow drift of activated aluminum atoms with the applied electric current [16]. Since the noise process and electromigration process are both strongly dependent on film microstructure, noise measurements may prove as an accurate predictor of interconnection reliability.

The values calculated for E_2 agree well with activation energies reported in the literature for electromigration. The statistical distribution of E_2 is currently being investigated. There is some preliminary data which suggests that very high values of current density will shift E_2 close to a bulk diffusion value. Such films may have few paths which support grain boundary diffusion in the direction of current flow [17].

Acknowledgments

The authors would like to thank Tom Crandell and the Analog Device Engineering Group at Harris Semiconductor, Melbourne, FL and Ken Rodbell at IBM-East Fishkill for fabricating the thin film samples used in the study. The authors also gratefully acknowledge the efforts of their graduate students L. M. Head and G. H. Massiha for their assistance in preparation of this paper. This work is partially supported by a grant from the National Science Foundation (ECS-8503522) and the Semiconductor Research Corporation.

References

[1] H. U. Schreiber, Solid State Elec., vol. 24, pp. 583-589, (1981).

[2] J. Spokas and C. Slichter, Phys. Rev., vol. 113, no. 6, pp. 1462-1472, (1959).

[3] I. A. Blech, E. Kinsbron, Thin Solid Films, vol. 25, pg. 327, (1975).

[4] H. U. Schreiber, Solid State Elec., vol. 28, no. 11, pp. 1153-1163, (1985).

[5] J. R. Black, Proc. of IEEE, vol. 57, pg. 1587, (1969).

[6] F. M. d'Heurle and P. S. Ho, "Electromigration in Thin Films" in *Thin Films; Interdiffusion and Reactions* J. Poate, K. N. Tu and J. Mayer, Eds., J. Wiley & Sons, N.Y., pg. 266, (1975).

[7] B. J. Root and T. Turner, Proc. of 23rd Intl. Rel. Phys. Symp., pp.100-107, (1985).

[8] C. C. Hong and D. L. Crook, Proc. of 23rd Intl. Rel. Phys. Symp., pp. 108-114, (1985).

[9] R. W. Pasco and J. A. Schwarz, Solid State Elec., vol. 26, pp. 445-452, (1983).

[10] J. M. Towner, Solid State Technol., vol. 27, pp. 287-289, (1984).

[11] J. L. Vossen, Appl. Phys. Lett., vol. 23, pp. 287-289, (1973).

[12] T. M. Chen, T. P. Djeu and R. D. Moore, Proc. of 23rd International Reliab. Phys. Symp., pp. 87-92, (1985).

[13] R. H. Koch, J. R. Lloyd and J. Cronin, Phys. Rev. Lett., vol. 55, no. 22, pp. 2487-2490, (1985).

[14] P. Dutta and P. M. Horn, Rev. Mod. Phys., vol. 53, pp. 497-516, (1981).

[15] J. G. Cottle,*Excess Noise and Its Relationship to Electromigration in Thin Film Interconnections*, Ph.D. Dissertation, University of South Florida, pg. 71, (1987).

[16] T. M. Chen, J. G. Cottle and L. M. Head, Proc. 9th Intl. Conf. on Noise in Phys. Systems, Montreal (1987).

[17] J. G. Cottle, T. M. Chen and K. P. Rodbell, Proc. of 26th International Reliab. Phys. Symp. (1988).

ELECTROMIGRATION IN THE PRESENCE OF A TEMPERATURE

GRADIENT: EXPERIMENTAL STUDY AND MODELLING

A. P. Schwarzenberger, C. A. Ross, J. E. Evetts and A. L. Greer

Department of Materials Science and Metallurgy, University of Cambridge,

Pembroke Street, Cambridge CB2 3QZ, UK.

ABSTRACT

Electromigration has reappeared as an important failure mechanism in integrated circuits as metallisation linewidths have decreased. Electromigration damage forms at sites of atom-flux divergence, of which the two main sources are microstructural inhomogeneities and local temperature variations (caused for example by track width changes or by nearby active circuits). We present a study of electromigration in sputter-deposited aluminium test structures with imposed temperature gradients using two test geometries. One enables the electromigration behaviour of the metallisation to be quickly characterised as a function of temperature, giving an activation energy; the other allows investigation of the changes in electromigration behaviour for different imposed temperature gradients. The results obtained are compared with computer modelling of the stress build-up during electromigration.

Microstructural Science for Thin Film
Metallizations in Electronics Applications
Edited by J. Sanchez, D.A. Smith and N. DeLanerolle
The Minerals, Metals & Materials Society, 1988

INTRODUCTION

Electromigration, the current-driven transport of the material of a conductor subjected to a high electric current density, has recently attracted intense study because it has been found to limit the further miniaturisation of integrated circuits (ICs). The metallisation in ICs, typically Al-Si-Cu, carries current densities of the order of 10^{10} Am^{-2}, and under these conditions, metal ions migrate in the direction of electron flow. This can lead to the formation of voids at points in a track from which material is depleted, or hillocks and whiskers at points where material accumulates. The metallisation finally fails when voids sever a track or when hillocks or whiskers cause short-circuits between tracks.

The average ionic flux J in a track undergoing electromigration can be expressed as [1]

$$J = \frac{N\rho Z^* e}{kT}(j - j_c)\frac{\delta}{d}D_o e^{-Q/kT} \tag{1}$$

where N is the ionic density, ρ is the resistivity, $Z^* e$ the effective charge on the ion (e is the electronic charge), k the Boltzmann constant, T the temperature, δ the grain boundary width, d the grain size, j the current density, j_c the threshold current density, D_o the grain boundary diffusion coefficient pre-exponential and Q the activation energy. In aluminium, transport is primarily via the grain boundaries, so grain boundary parameters, such as diffusivity, have been used. It is important to note that electromigration damage (hillocks and voids) can only form where there is a divergence in the electromigration flux J, caused by variations in any of the parameters on the right-hand side of (1). For instance, if the grain size of the metallisation changes, perhaps due to a change in the substrate, the flux-carrying capacity of the track is altered and there is a flux divergence. Cross-section changes in themselves do not lead to flux divergences, because a reduction in cross-sectional area leads to a higher current density in the remaining section and thus a higher flux. However, a section change can cause a change in the local self-heating and therefore a temperature-induced divergence.

Temperature changes are important sources of flux divergence, since the flux depends exponentially on temperature. For example, at 200°C in aluminium, a 5°C change in temperature results in a change of more than 10% in the electromigration flux. Temperature changes may arise from changes in the thermal properties of the substrate, as the metallisation passes over other features in the substrate, or by changes in the self-heating caused by section changes, for instance at steps over substrate features. Temperature changes can also lead to thermomigration (transport of material in a temperature gradient), though this is small for aluminium tracks in ICs as compared to electromigration. We will present a study of electromigration in structures with intentional temperature gradients, and will show how temperature-induced flux divergences lead to damage formation as predicted by a model for electromigration damage [2].

EXPERIMENTAL WORK

Our experimental work has involved two different test geometries. The first geometry uses the drift stripe technique pioneered by Blech [3,4,5] combined with an imposed temperature gradient, allowing an activation energy for the electromigration drift velocity to be found from a single experiment. In the second technique, a metallisation track with artificially-introduced voids along its length is current-stressed while a temperature gradient is maintained along it using an on-chip heater track. By this approach we can study the effect of controlled temperature-induced flux divergences on electromigration damage formation.

(a) Drift velocity measurement

The average drift velocity of the migrating ions during electromigration is given by $v = J/N$. This velocity has been measured directly [3,4,5] in aluminium and gold, using a geometry consisting of blocks of test metallisation on a refractory underlayer. A current in the underlayer is picked up by the test metallisation, and the cathode end of the block drifts at velocity v in the direction of current flow (Figure 1). The velocity is measured using optical microscopy.

In the present work, the activation energy for drift was measured from the variation of v with temperature. This was done for aluminium in a single test, in which a temperature gradient was imposed along a niobium refractory track on which blocks of aluminium drifted. Samples were prepared from bilayer films of pure aluminium on niobium, deposited by UHV getter sputter deposition onto 12mm x 3mm x 0.5 mm sapphire substrates, and fabricated into finished devices using optical lithography as has been described previously [6]. The thickness of the individual metal layers was the same, and the total film thickness was chosen between 0.2μm and 1.0μm. The films were annealed at 450°C in argon before fabrication to promote grain growth. X-ray diffraction and resistance ratios to 77K were used to characterise the film microstructure, and it was found that annealing caused the as-deposited fibrous structure to develop into an equiaxed structure, with grains of a similar size to the film thickness. For example, one 0.35μm thick aluminium film showed a resistance ratio change from 3.09 to 3.63 on annealing.

Each sample consisted of an 8000μm x 20μm niobium underlayer track, along which there were 36 200μm x nominally 20μm aluminium stripes separated by 20μm spaces. A temperature difference of up to 50°C could be maintained between the ends of the stripe by using miniature thermopiles. The temperature at eight points along the track was found from the four-terminal resistance of meander resistors present on the sample, which were calibrated from the resistance ratios of the annealed films. The temperatures were also measured using a thermocouple. The overall temperature was controlled by mounting the sample on a heated copper plate. The apparatus is illustrated in Figure 2.

The results of a preliminary test on a 0.45μm aluminium on 0.5μm niobium film are shown in Figure 3. The sample was tested at a current density of 0.9 10^{10} Am^{-2} for 840 minutes, and the temperature ranged between 135°C and 173°C along the length of the test track. The temperature variation along the track is given in Figure 3(a); self-heating has raised the overall temperature and has slightly altered the shape of the temperature profile.

The resistivity of the metallisation rises linearly with temperature for small temperature changes, so the $1/T$ term in the flux is approximately cancelled and the temperature dependence of the flux therefore arises just from the exponential term [see Eqn. 1]. Variations in Z^* and j_c with temperature are small compared to the change in diffusivity. The stripe length of 200μm corresponds to a j_c value of 1 10^9 Am^{-2} [6], which is much less than the applied current density, so threshold effects have been neglected. To take account of slight width variations, the drifted area A was measured for each stripe. From Eqn. 1, this is given by

$$A = \gamma \exp(-Q/kT) \tag{2}$$

where γ is a constant. Figure 3(b) shows a plot of $\log A$ vs $1/T$, from which a value of $Q = 0.73 \pm 0.03$ eV is obtained. Deviations from linearity in this plot are caused by local fluctuations in the track width, introduced during fabrication, because there is more Joule heating in the narrower regions of the track. Peaks in the plot are found to correspond to narrow regions of the track, which emphasizes the importance of well-controlled sample fabrication techniques. The value of Q was derived from aluminium stripes whose widths were between 14.9 and 16.6μm whereas the widths in the sample varied between 12.2 and 16.7μm.

Comparison with results obtained by a drift-stripe technique [7] indicate that this particular value for Q is high for the expected grain-boundary transport, which is typically 0.4 - 0.5 eV. The activation energy derived from electromigration tests will differ from the activation energy for grain boundary self-diffusion because Z^* and j_c have a small temperature dependence. Further experiments are in progress to investigate the temperature dependence of drift more fully and to refine the present techique, to improve temperature control and reduce self-heating. The value for $\delta Z^* D_o \exp(-Q/kT)$ at 150° is 6 10^{-22} m^3 s^{-1}, which is within the experimental range quoted in a review article [1].

179

Although still at a very preliminary stage, this technique is valuable because it allows thin film metallisation to be characterised in a single test, in which, for example, the deposition conditions of the film are constant. We are at present developing a method in which current density and temperature can both be varied simultaneously so that a matrix of experimental conditions can be studied. This will enable rapid characterisation of different types of metallisation to be made, and their suitability for integrated circuit metallisation examined.

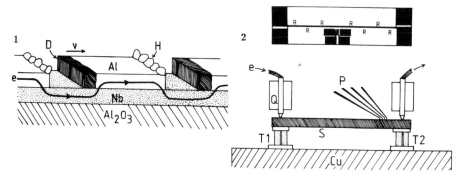

Figure 1: Drift stripe geometry. The electron flow is represented by e. The current is picked up by the aluminium which drifts at velocity v, and the drifted region, from which material has been depleted, is marked D. Material accumulates as hillocks, H, at the anode end of the stripe.

Figure 2: Diagram of drift velocity apparatus. The 12mm sample S is clamped to two thermopiles $T1$ and $T2$ resting on a copper hotplate by two sprung contacts, Q, through which the current e is introduced. One thermopile is driven in reverse and one forwards, to produce a temperature gradient. The resistance of four-terminal resistors R is measured by 5μm-tipped probes P. A plan of the sample is shown in the upper diagram, in which the position of the resistors is shown and one is drawn fully. In the plan, the row of drift stripes runs horizontally between the resistors, and is connected to four large pads at the corners of the sample.

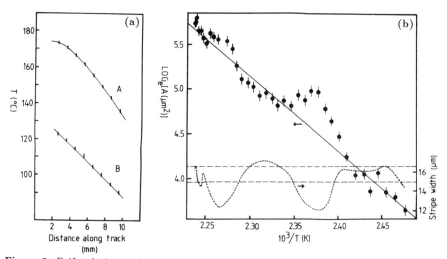

Figure 3: Drift velocity results.
(a) The temperature profile in the sample, powered A (*i.e.* with the test current flowing) and unpowered B (*i.e.* the temperature profile just from the hotplate and thermopiles).
(b) Plot of logarithm of drifted area *vs.* inverse temperature. The dotted line shows the width along the track.

180

(b) Temperature induced divergences

The strong dependence of the electromigration flux on temperature means that small temperature variations on an active integrated circuit will give rise to large variations in the flux, i.e. flux divergences. However, the actual temperature at any point on a metallisation track is not often well characterised, despite the fact that these divergences may well be the most significant ones present, and hence the cause of failure. In this work we have designed and fabricated special test structures which allow temperature gradients to be both imposed and monitored along the length of a metallisation track and present some preliminary results.

The test structure is shown in plan and cross section in Figure 4. Samples were fabricated on 6-inch silicon wafers on which 0.1μm of thermal oxide had been grown. The test structure consists of a 0.7μm thick by 3.5μm wide by 540μm long aluminium test track which is electrically isolated from a 0.9μm thick tungsten underlayer by 0.2μm of oxide. The aluminium test tracks have artificial voids (diameter 1.5μm) introduced along their length for two reasons: firstly to probe the local conditions in the metallisation track by providing a site for damage formation and secondly because they are of interest in a continuing study of void migration [8]. The tungsten was deposited by plasma-enhanced CVD (renucleated), and patterned. The aluminium was sputter deposited and annealed for 30 minutes at 450°C after patterning. The patterning was by standard optical lithography and dry-etching techniques. The thicknesses were characterised using a Talysurf profilometer. The tungsten was patterned to provide the central heater stripe ('H' in Figure 4) and the five temperature sensors (e.g. 'T'). Redundant tungsten ('R') is left under the entire length of the aluminium test track ('A') to eliminate complications which would arise from steps in that test track. Samples were diced out of the wafer with the heater track in the centre so that a symmetrical temperature profile around the heater track is obtained. The samples were packaged in ceramic dual-in-line packages which were heat-sunk to a water cooled copper block during testing, electrical connection being made by aluminium wire-bonds. Temperature measurements are made by four terminal resistance measurements of both the five temperature sensors and the heater track. These were calibrated by the resistance ratio of the tungsten. The use of multimeters with unearthed inputs eliminates electrical interference between the temperature measurements and the current in the heater track. The temperature profile in the sample is expected to be symmetrical about the heater track, so the width of the temperature sensors should be kept to a minimum, but their length is not critical. This technique should allow the temperature of each of these sensors to be measured to within ± 1°C, although an accuracy of ± 2.5°C seems more reasonable in practice. The temperature gradient along the test track can be controlled by varying the amount of heat sinking through the package: greater heat sinking giving a steeper temperature gradient.

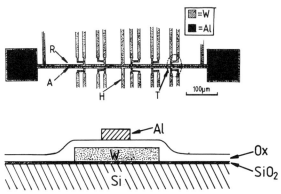

Figure 4: Plan view (top) and cross-section (bottom) of test structure designed for electromigration testing with an imposed temperature gradient along the length of the test track. See text for details.

An example of an extreme temperature profile obtained is shown in Figure 5, having a temperature difference of about 50°C between the centre and ends of the test track. This is an average temperature gradient of about 1°C per 4μm, with a maximum temperature gradient of up to double this near to the heater track. The experiments described below used a more typical temperature difference of about 20°C between the centre and ends of the track.

These temperature gradients are similar to, or less severe, than those found in commercial IC's where a hot-spot may be more than 10°C hotter than nearby regions of the microcircuit. This design and method can therefore be usefully used to study in a test environment the effect of temperature gradients on IC metallisation failure mechanisms.

A one-dimensional model which predicts the build up of stress in test tracks for a given set of flux-divergences has been previously described [2]. This has been used here to compare the expected build up of stress in two metallisation tracks: the first with a few degrees self-heating along the entire length of the test track, and the second with an imposed temperature gradient similar to that shown in Figure 5. The results of the computer simulation are shown in Figure 6(a) and 6(b) respectively. For the self- heated test track (Figure 6a) flux divergences are found at the ends of the stripe near to the cooler bond pads, so the peaks of compressive (positive) and tensile (negative) stress are towards the ends of the test track. Figure 6b shows the temperature profile along a test track with imposed heating. We make the simplifying assumption that heat flow occurs cylindrically outwards from the heater track in the thick silicon and ignore heat flow in the other thin layers. The peaks of stress are now found to be much nearer to the centre of the test track. So we expect the main sites of electromigration damage formation to move depending on the gradient of the imposed heating.

Figure 7 shows experimental observations of the difference between self- heated and imposed-heating test tracks. Figure 7a shows the centre of a self-heated test track. No electromigration damage can be seen, but, if we look 95μm either side of the heater track, we find on one side the site where failure by voiding has occured (Figure 7b) and on the other side some small hillocks (Figure 7c). However, the test track with imposed heating shows rather different behaviour, with the electromigration damage concentrated near to the heater track (Figure 7d). So we see that in a nominally uniform test track, the sites of electromigration damage are heavily influenced by the temperature gradient present.

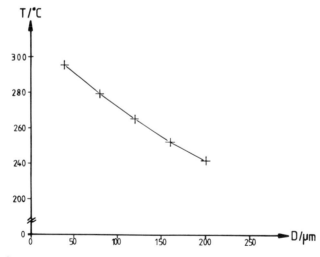

Figure 5: Plot of temperature T against distance D away from the heater track in a test structure as in figure 4. The bond pad is 270μm away from the heater track.

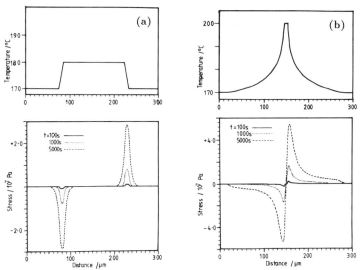

Figure 6: Computer simulations of the stress build up in an aluminium test track for two different temperature profiles, both with a base temperature of 170°C and the electron current from left to right. Positive stress is compressive.

(a) Test track at a uniform 10°C hotter than the bond pads.

(b) Centre of test track (i.e. at the heater track) 30°C hotter than the bond pads. The profile is obtained by assuming that all the heat flow occurs cylindrically outwards from the heater track in the thick silicon.

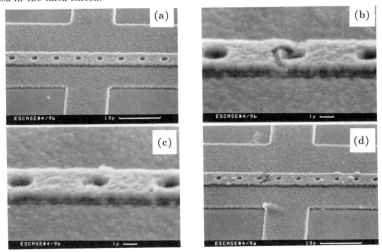

Figure 7: Scanning electron micrographs of Al test tracks. The samples were tested at $2\ 10^{10}$ Am^{-2} for about 40 minutes at a base temperature of 200°C (i.e. before imposed or self heating). See text for further details

(a) Central region of a self heated test track.

(b) Detail 95μm to left of figure 7a.

(c) Detail 95μm to right of figure 7a.

(d) Central region of a test track that had a temperature gradient imposed by the tungsten heater track. Voids are visible to the left of the heater track and hillocks to the right.

183

CONCLUSIONS

The experimental techniques presented here demonstrate two ways of using temperature gradients to characterise the electromigration behaviour of thin film metallisation. The drift-stripe technique allows the activation energy for electromigration drift to be assessed in a single test. The use of an embedded heater strip under a metallisation track to cause flux divergences has been shown to give a pattern of electromigration damage consistent with the results of a computer simulation. This work demonstrates the importance of temperature gradients as sources of electromigration flux divergence in IC metallisation and therefore the necessity of controlling the temperature distribution in an active IC in order to optimise its lifetime.

ACKNOWLEDGEMENTS

This work was supported by the Science and Engineering Research Council, Plessey Research Caswell Ltd and the General Electric Company Hirst Research Centre. CAR would like to thank R.E. Somekh for film deposition, and APS would like to thank Plessey Research for help with sample fabrication. We thank J. Sanchez for helpful discussions.

REFERENCES

1. F.M. d'Heurle and P.S. Ho, *"Electromigration in thin films"*, Chapter 8 of *Thin Films - Interdiffusion and Reactions*, ed. J.M. Poate, K.N. Tu and J.W. Mayer (John Wiley and Sons Inc., New York, 1978)

2. C.A. Ross and J.E. Evetts, *"A model for electromigration damage in terms of flux divergences"*, *Scripta Metall.* **21**(8) 1077 (1987)

3. I.A. Blech and E. Kinsbron, *"Electromigration in thin gold films on molybdenum surfaces"*, *Thin Solid Films* **25** 327 (1975)

4. I.A. Blech, *"Electromigration in thin aluminium films on titanium nitride"*, *J. Appl. Phys* **47**(4) 1203 (1976) and *Erratum, J. Appl. Phys* **48**(6) 2648 (1977)

5. I.A. Blech and Conyers Herring, *"Stress generation by electromigration"*, *App. Phys. Lett.* **29**(3) 131 (1976)

6. C.A. Ross and J.E. Evetts, *"A study of threshold and incubation behaviour during electromigration in thin film metallisation"*, presented at the 1987 MRS Fall Meeting, Boston, MA, 1987 (published in Proceedings, in press)

7. H.-U. Schreiber, *"Activation energies for the different electromigration mechanisms in aluminium"*, *Sol. St. Electron.* **24** 583 (1981)

8. A.P. Schwarzenberger and A.L. Greer, *"Asymmetries in the formation of electromigration damage around divergence dipoles in a metallization track"*, presented at the 1987 MRS Fall Meeting, Boston, MA, 1987 (published in Proceedings, in press)

MICROSTRUCTURAL DYMANICS IN ALUMINUM THIN FILMS

AND RELATED RELIABILITY ISSUES

C. Y. Wong, S. S. Iyer

IBM T. J. Watson Research Center
Yorktown Heights, New York 10598

Abstract

Aluminum and aluminum-based films are widely used as the interconnect metallurgy in most integrated circuit technologies. The microstructure of these polycrystalline films plays a very important role in defining both the properties of the film as well as their processibility. While it has been shown that longer lifetimes correlate to larger grain size, it has also been shown that such increases in lifetime are associated with concomitant increases in the spread of the failure distribution. This spread is linked, to a large extent, to the nonuniformity in the grain size distribution. In this talk, we will discuss the grain growth behavior of pure aluminum films under a variety of processing conditions for both blanket and patterned films. Implications of our results on the reliability and processibility of the metal films will also be discussed.

Microstructural Science for Thin Film
Metallizations in Electronics Applications
Edited by J. Sanchez, D.A. Smith and N. DeLanerolle
The Minerals, Metals & Materials Society, 1988

MICROSTRUCTURAL EFFECTS ON ELECTROMIGRATION RESISTANCE

IN THIN FILM INTERCONNECTS

Thomas Kwok

IBM T. J. Watson Research Center
Yorktown Heights, New York 10598

Abstract

It is generally believed that there is a correlation between microstructure and electromigration resistance, especially in submicron metal lines where the linewidth is comparable to the grain size. In the first part of this paper, the present understanding on the effects of microstructure on electromigration resistance in fine metal lines will be reviewed. This paper also investigates the effects of grain size and grain morphology on electromigration lifetime in Al-Cu fine lines. By annealing 2.0 μm wide Al-Cu lines at elevated temperatures, which induces grain growth, the electromigration lifetime is found to increase with the grain size. However, in Al-Cu submicron lines, where the grain size is comparable to the linewidth and film thickness, the grain structure is found to resemble a bamboo structure. These submicron metal lines are also found to have longer electromigration lifetime than other fine metal lines. So the grain morphology becomes an important factor affecting the electromigration resistance in submicron metal lines. The microstructural modifications, such as columnar and bamboo structures, could explain the improvement of electromigration resistance in submicron metal lines. The effects of film thickness and linewidth on the microstructural modifications in the submicron metal lines are discussed.

Microstructural Science for Thin Film
Metallizations in Electronics Applications
Edited by J. Sanchez, D.A. Smith and N. DeLanerolle
The Minerals, Metals & Materials Society, 1988

SUBJECT INDEX

AUTHOR INDEX